PEARSON EDEXCEL INTERNATIONAL GCSE (9–1)

GEOGRAPHY

Student Book

Michael Witherick

Published by Pearson Education Limited, 80 Strand, London, WC2R 0RL.

https://www.pearson.com/international-schools

Copies of official specifications for all Edexcel qualifications may be found on the website: https://qualifications.pearson.com

Text © Pearson Education Limited 2017
Edited by Elizabeth Barker
Designed by Cobalt id
Typeset by Cobalt id
Original illustrations © Pearson Education Limited 2017
Illustrated by © Cobalt id
Cover design by Pearson Education Limited
Cover photo/illustration © Getty images: wiratgasem

Inside front cover: [Shutterstock.com/Dmitry Lobanov]

The rights of Michael Witherick to be identified as the author of this work have been asserted by him in accordance with the Copyright, Designs and Patents Act 1988.

First published 2017

25 24 23 22 21
10 9 8 7

British Library Cataloguing in Publication Data
A catalogue record for this book is available from the British Library

ISBN 978 0 435184 83 4

Printed by Neografia in Slovakia

Endorsement Statement
In order to ensure that this resource offers high-quality support for the associated Pearson qualification, it has been through a review process by the awarding body. This process confirms that this resource fully covers the teaching and learning content of the specification or part of a specification at which it is aimed. It also confirms that it demonstrates an appropriate balance between the development of subject skills, knowledge and understanding, in addition to preparation for assessment.

Endorsement does not cover any guidance on assessment activities or processes (e.g. practice questions or advice on how to answer assessment questions) included in the resource, nor does it prescribe any particular approach to the teaching or delivery of a related course.

While the publishers have made every attempt to ensure that advice on the qualification and its assessment is accurate, the official specification and associated assessment guidance materials are the only authoritative source of information and should always be referred to for definitive guidance.

Pearson examiners have not contributed to any sections in this resource relevant to examination papers for which they have responsibility.

Examiners will not use endorsed resources as a source of material for any assessment set by Pearson. Endorsement of a resource does not mean that the resource is required to achieve this Pearson qualification, nor does it mean that it is the only suitable material available to support the qualification, and any resource lists produced by the awarding body shall include this and other appropriate resources.

Acknowledgements

We are grateful to the following for permission to reproduce copyright material:

Figures
Figures 1.18 and 1.20 © Crown Copyright, contain public sector information licensed under the Open Government Licence (OGL) v3.0. http://www.nation-alarchives.gov.uk/doc/open-government-licence/version/3/; Figure 2.36 from http://www.grida.no/resources/7035, Delphine Digout © UNEP/GRID-Arendal; Figure 3.3 from *GCSE Geography AQA A*, Oxford University Press (Rowles, N., Holmes, D. and Digby, B. 2016) p. 26, edited by Simon Ross. By permission of Oxford University Press, UK; Figure 3.14 adapted from Oxfam America, © 2016 Oxfam America Inc. All rights reserved; Figure 3.15 adapted from Why the earthquake in Italy was so destructive, *The Washington Post*, Sarah Kaplan, 24 August 2016. Map: Tim Meko and Ashley Wu, permission conveyed through Copyright Clearance Center, Inc.; Figure 3.16 from A year after the 2010 Merapi eruption: volcano hazard and Indonesian government mitigation measures, Sri Hidayati, Surono and Subandriyo, Center for Volcanology and Geological Hazard Mitigation (CVGHM), Indonesia, www.http://miavita.brgm.fr; Figure 3.38 from The human impact of tropical cyclones: a historical review of events 1980-2009 and systematic literature review (Doocy, S., Dick, A., Daniels, A. and Kirsch, T.D.), PLOS Currents Disasters, 16 April 2013; Figure 3.39 from Estimating tropical cyclone damages under climate change in the Southern Hemisphere using reported damages, *Environmental and Resource Economics*, Vol. 58 (3), pp. 473¬¬–490 (Seo, S.N. 2014), Copyright © 2013, Springer Science+Business Media Dordrecht. With permission of Springer; Figure 4.20 from *Contemporary Case Studies: Population and Migration*, Philip Allan (Witherick, M. 2004) Figure 28, © Philip Allan Updates 2004. Reproduced by permission of Philip Allan (for Hodder Education); Figure 4.22 from *International Energy Outlook 2013*, EIA; Figure 4.25 from International Energy Agency, © OECD/IEA 2017, www.iea.org/t&c; Figure 4.26 from mapsofworld.com, Copyright 2014 'MapsOfWorld'. All rights reserved; Figure 4.29 from Energy Information Administration; Figure 4.30 from *World Energy Outlook Special Report: India Energy Outlook*, IEA Publishing (2015) p. 22, Figure 1.3, © OECD/IEA 2015 Licence: www.iea.org/t&c; as modified by Pearson Education; Figure 4.31 GHI Solar Map © 2017 Solargis; Figure 5.7 adapted from Deforestation in the Amazon, Rhett Butler (2009), mongabay.com, Copyright Rhett Butler 2009; Figure 6.24 from *Contemporary Case Studies: Cities and Urbanisation*, Philip Allan (Witherick, M. and Adams, K. 2006), Figure 52, © Philip Allan Updates 2006. Reproduced by permission of Philip Allan (for Hodder Education); Figure 7.2 from *Living Planet Report 2000*, WWF - World Wide Fund for Nature (formerly World Wildlife Fund) Gland, Switzerland p. 11, Map 4, © text 2000 WWF. All rights reserved; Figure 7.5 from Millennium Ecosystem Assessment; Figure 7.20 from mongabay.com; Figure 7.28 © US EPA; Figure 7.34 from The great flood of London: Experts warn risk to Capital from rising sea levels 'worse than feared', *Mail Online*, Nick McDermott, 15 May 2013; Figure 8.4 from The WTO - the global 'umpire' of fair trade, Ingram Pinn, Financial Times, 8 August 2012. Used under licence from the Financial Times. All rights reserved; Figure 8.5 © CIA; Figure 8.16 from What is a single market? Definition and meaning, Market Business News; Figure 9.7 from *Development, Disparity and Dependence: A Study of the Asian Pacific Region (EPICS)*, Stanley Thornes (Witherick, M. 1998); Figure 9.14 from Global Peace Index 2014, Vision of Humanity © Institute for Economics and Peace; Figure 9.23 from A tale of two economies, *The Economist*, 16 May 2015, Copyright © The Economist Newspaper Limited 2015. All rights reserved; Figure 9.24 from *Edexcel GCSE Geography A*, Edexcel (Yates, N., Palmer, A., Witherick, M. and Wood, P. 2012) p. 181, Figure 6, Pearson Education Ltd; Figure 9.25 from *The World Factbook*, Central Intelligence Agency; Figure 9.31 from High-speed rail opponents 'portrayed as posh nimbys' by peer's lobbying firm *The Observer*, Jamie Doward, 6 April 2013, Copyright Guardian News & Media Ltd. 2017, www.theguardian.com

SECTION A: PHYSICAL ENVIRONMENTS

SECTION B: HUMAN ENVIRONMENTS

ABOUT THIS BOOK

This book is written for students following the Pearson Edexcel International GCSE (9–1) Geography specification.

This book has been structured so that it can be used in order, both in the classroom and for independent learning. It explains and illustrates all the subject content of the specification; it is on the knowledge and understanding of this content that the student will be examined.

Each chapter has a short introduction to help students start thinking about the topic. It also sets out the learning objectives which will be covered by the end of the chapter.

Throughout the book there are margin boxes setting a range of different tasks. All of them are aimed at helping master the content and skills required.

In their study of the nine topics, students have to include at least two countries as case studies. One should be a developed country; the other either an emerging or a

developing country. In this book, all options are covered so that in each chapter there are at least three national case studies that cover the three main levels of economic development. An important aspect of these country case studies is comparison. Those comparisons should be between the developed country and either the emerging or developing country selected.

The boxes in the margins and text do one of several things:
1 Ask short questions to check knowledge and understanding (**Check your understanding**)
2 Suggest a range of activities that build on what has been read in the chapter, from issues for discussion to exercises based on a figure or table (**Activity** and **Skills**).
3 Give you extra information or help (**Did you know?**).

There are also prompt boxes in Sections A and B with advice on making use of fieldwork. Fieldwork can be found on the ActiveBook.

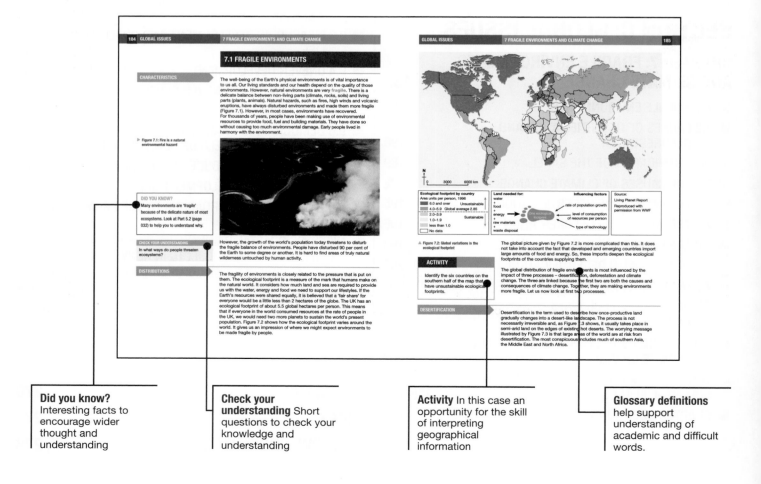

Did you know? Interesting facts to encourage wider thought and understanding

Check your understanding Short questions to check your knowledge and understanding

Activity In this case an opportunity for the skill of interpreting geographical information

Glossary definitions help support understanding of academic and difficult words.

END OF CHAPTER QUESTIONS

There are three items here:
- a checklist of integrated skills recommended in the specification for the topic
- review questions using approved exam command words
- short and longer exam-style practice questions.

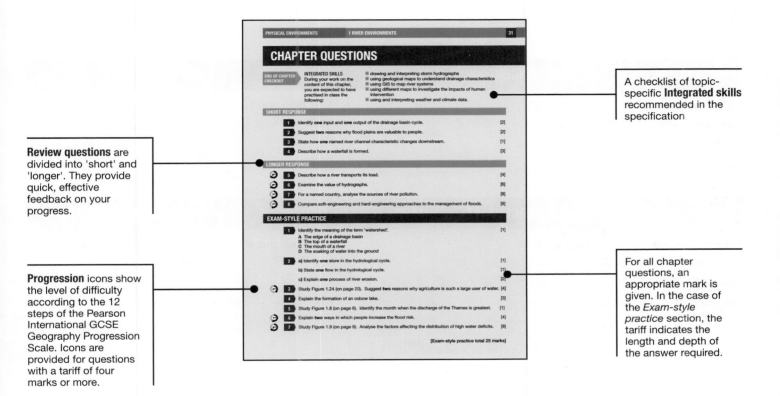

A checklist of topic-specific **Integrated skills** recommended in the specification

Review questions are divided into 'short' and 'longer'. They provide quick, effective feedback on your progress.

Progression icons show the level of difficulty according to the 12 steps of the Pearson International GCSE Geography Progression Scale. Icons are provided for questions with a tariff of four marks or more.

For all chapter questions, an appropriate mark is given. In the case of the *Exam-style practice* section, the tariff indicates the length and depth of the answer required.

THE ACTIVEBOOK

The ActiveBook contains not only exam-type questions, but also:
- advice about the fieldwork required; questions about fieldwork will be set in Section B of both examination papers
- extra case studies for the following chapters: Chapter 1, Chapter 2 (x2), Chapter 6 (x2), Chapter 7 (x2), Chapter 8 (x4) and Chapter 9.

ASSESSMENT OVERVIEW

The following tables give an overview of the assessment for the Pearson Edexcel International GCSE (9–1) in Geography.

We recommend that you study this information closely to help ensure that you are fully prepared for this course and know exactly what to expect in the assessment.

PAPER 1	PERCENTAGE	MARK	TIME	AVAILABILITY
Physical geography Written examination paper Paper code 4GE1/01 Externally set and assessed by Edexcel	40%	70	1 hour and 10 minutes	June examination series First assessment June 2019

PAPER 2	PERCENTAGE	MARK	TIME	AVAILABILITY
Human geography Written examination paper Paper code 4GE1/02 Externally set and assessed by Edexcel	60%	105	1 hour and 45 minutes	June examination series First assessment June 2019

ASSESSMENT OBJECTIVES AND WEIGHTINGS

ASSESSMENT OBJECTIVE	DESCRIPTION	% IN INTERNATIONAL GCSE
AO1	Demonstrate knowledge of locations, places, processes, environments and different scales.	15–16
AO2	Demonstrate geographical understanding of: • concepts and how they are used in relation to places, environments and processes • the interrelationships between places, environments and processes.	25–26
AO3	Apply knowledge and understanding to interpret, analyse and evaluate geographical information and issues and to make judgements.	34–35 (approx. 13% applied to fieldwork context/s)
AO4	Select, adapt and use a variety of skills and techniques to investigate questions and issues and communicate findings.	24–25 (approx. 10% used to respond to fieldwork data and context/s)

RELATIONSHIP OF ASSESSMENT OBJECTIVES TO UNITS

UNIT NUMBER	ASSESSMENT OBJECTIVE			
	AO1	AO2	AO3	AO4
Paper 1	7.1%	12.9%	17.9%	12.1%
Paper 2	8.5%	12.9%	16.2%	12.4%
Total for International GCSE	15–16%	25–26%	34–35%	24–25%

PAPER 1	DESCRIPTION	MARKS
Physical geography Paper code 4GE1/01	**Structure** Paper 1 assesses 40% of the total Geography qualification. There will be two sections on the paper.	Total number of marks available: 70
	Section A This section consists of multiple-choice, short-answer, data-response, and open-ended questions. Students choose two out of three questions on: river environments, coastal environments, and hazardous environments.	50
	Section B This section requires students to use knowledge and understanding from research and fieldwork that they have carried out. Students are not allowed to take materials into the examination. Students choose one out of three fieldwork-related questions on: river environments, coastal environments, and hazardous environments.	20

PAPER 2	DESCRIPTION	MARKS
Physical geography Paper code 4GE1/02	**Structure** Paper 2 assesses 60% of the total Geography qualification. There will be three sections on the paper.	Total number of marks available: 105
	Section A This section consists of multiple-choice, short-answer, data-response, and open-ended questions. Students choose two out of three questions on: economic activity and energy, rural environments, and urban environments.	50

Section B This section requires students to use knowledge and understanding from research and fieldwork that they have carried out. Students are not allowed to take materials into the examination. Students choose one out of three fieldwork-related questions on: economic activity and energy, rural environments, and urban environments.	20
Section C This section consists of multiple-choice, short-answer, data-response, and open-ended questions. Students choose one out of three questions on: fragile environments and climate change, globalisation and migration, and development and human welfare.	35

EXAM COMMAND WORD DEFINITIONS

This table lists the command words that may be used in the examinations for this qualification and their definitions. It is important that you understand what the question is asking, in order to answer and receive full marks.

COMMAND WORD	DEFINITION
Identify/state/name	Recall or select one or more pieces of information.
Define	State the meaning of a term.
Calculate	Produce a numerical answer, showing relevant working.
Label	Add a label/labels to a given resource, graphic or image.
Draw/plot	Create a graphical representation of geographical information.
Compare	Find the similarities and differences between two elements given in a question. Each response must relate to both elements, and must include a statement of their similarity/difference.
Describe	Give an account of the main characteristics of something or the steps in a process. Statements in the response should be developed but do not need to include a justification or reason.
Explain	Provide a reasoned explanation of how or why something occurs. An explanation requires a justification/exemplification of a point. Some questions will require the use of annotated diagrams to support explanation.
Suggest	Apply understanding to provide a reasoned explanation of how or why something may occur. A suggested explanation requires a justification/exemplification of a point.

Examine	Break something down into individual components/processes; say how each one individually contributes to the question's theme/topic and how the components/processes work together and interrelate.
Assess	Use evidence to determine the relative significance of something. Give consideration to all factors and identify which are the most important.
Analyse	Investigate an issue by breaking it down into individual components; make logical, evidence-based connections about the causes and effects or interrelationships between the components.
Evaluate	Measure the value or success of something and ultimately provide a substantiated judgement/conclusion. Review information and then bring it together to form a conclusion, drawing on evidence such as strengths, weaknesses, alternatives and relevant data.
Discuss	Explore the strengths and weaknesses of different sides of an issue/question. Investigate the issue by reasoning or argument.

1 RIVER ENVIRONMENTS

LEARNING OBJECTIVES

By the end of this chapter, you should know:

- How water moves through the hydrological cycle

- The main features of a drainage basin

- The factors affecting river regimes

- The fluvial processes involved in the formation of river channels and valleys

- How the characteristics of river channels and valleys change along the course of a named river

- The difference between upland and lowland river landscapes

- The rising demand for water and the creation of areas of water shortage and water surplus

- The importance of water quality; the storage and supply of clean water

- The causes of river flooding, and how to predict and prevent it

This chapter is about rivers. They are a vital part of the global circulation of water. Rivers are responsible for the creation of landforms found throughout the world. They are valuable to us because they supply much of the water we use in our everyday lives. At times, because of the risk of flooding, they can become hazards that threaten people and their settlements.

1.1 THE HYDROLOGICAL CYCLE – A CLOSED SYSTEM

Fresh water is essential for life on Earth. This water is constantly being recycled as it moves through what is called the hydrological cycle. This is a global circulation of water and it is a giant **closed system** (Figure 1.1). This means that there is a fixed amount of water because water neither enters nor leaves the Earth and its atmosphere.

▼ Figure 1.1: Stores and flows in the hydrological cycle

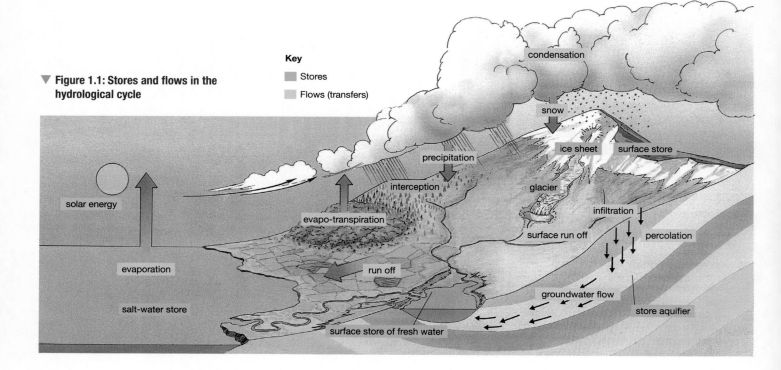

Key
- Stores
- Flows (transfers)

condensation

snow

ice sheet   surface store

precipitation

interception

glacier

infiltration

solar energy

evapo-transpiration

surface run off

percolation

evaporation

run off

groundwater flow

store aquifer

salt-water store

surface store of fresh water

STORES

During the hydrological cycle, water is held in a number of **stores** and then moves between them by means of a series of flows, also called **transfers**. The stores in the cycle are as follows.

CHECK YOUR UNDERSTANDING

Which store do you think holds the largest amount of water?

- The atmosphere – where the water exists either as water vapour or as minute droplets in clouds.
- The land – where water is stored on the surface in rivers, lakes and **reservoirs**. Water is taken in by plants and stored in vegetation for short periods of time. It is also stored below ground in the bedrock. This is known as the **groundwater** store (see **aquifer**). Water mostly exists in these stores in a liquid form. However, it can also exist in a solid form as snow and ice, for example in ice sheets, glaciers and snowfields.
- The sea – it is estimated that over 95 per cent of the Earth's water is stored in the sea. This is mostly held in liquid form, but also as ice, for example the icebergs in high-latitude seas.

SKILLS REASONING

ACTIVITY

List the possible physical consequences of melting ice sheets and glaciers.

While the amount of water in the global hydrological cycle cannot change, the proportion held in the different stores can. These variations are caused by changes in the Sun's energy. For example, an increase in the Sun's energy will lead to more evaporation and possibly to the melting of ice sheets and glaciers.

FLOWS (TRANSFERS)

The transfers of water that take place between stores do so through a variety of **flows** as listed below.

- Evaporation – the hydrological cycle starts with evaporation due to the heat of the Sun. Water is converted from a liquid into a gas (called water vapour). This takes place from the surface of the sea and from water surfaces (ponds and lakes) on land. Evaporation is particularly important in the transfer of water from the sea store into the atmosphere.
- Transpiration – plants take up liquid water from the soil and 'breathe' it into atmosphere as water vapour.
- **Evapotranspiration** – the loss of moisture from the ground by direct evaporation from water bodies and the soil, plus transpiration from plants.
- Condensation – the change in the atmosphere when water vapour cools and becomes liquid. The liquid takes the form of water droplets that appear in the atmosphere as clouds.
- Precipitation – the transfer of water in any form (rain, hail or snow) from the atmosphere to the land or sea surface.
- Overland flow – most precipitation that hits the ground moves due to gravity and eventually enters a stream, river or lake. This is known as run off.
- Infiltration and percolation – the transfer of water downwards through the soil and rock into the aquifer or groundwater store.
- Throughflow – this takes place between the ground surface and the top of the groundwater store. As a result of gravity, water moves slowly through the soil until it reaches a stream or river.
- Groundwater flow – this happens in the rocks of the aquifer and is the underground transfer of water to rivers, lakes and the sea.

All these transfers come together to form a circle involving the three major stores (the atmosphere, the land and the sea). We might imagine that the cycle starts and finishes in the sea. However, some of the water that falls as rain on the land may never reach the sea. Instead, it is returned directly back to the atmosphere from the land by the transpiration of plants and evaporation from both soil and water bodies

CHECK YOUR UNDERSTANDING

What is the difference between a store and a flow?

1.2 DRAINAGE BASINS AND THEIR FEATURES

Every river has its own drainage basin or catchment area. Each drainage basin is a system. There is a movement of water within it that is rather like a small-scale hydrological cycle. The drainage basin cycle involves stores and flows (Figure 1.2). However, an important difference between the drainage basin and the hydrological cycle is that the drainage basin is an **open system** (Figure 1.3). A drainage basin has external inputs and outputs. The amount of water in the basin system varies over time. In the hydrological cycle the amount of water remains exactly the same.

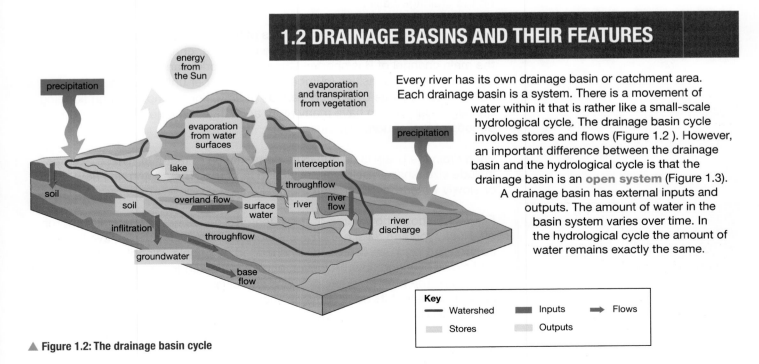

▲ Figure 1.2: The drainage basin cycle

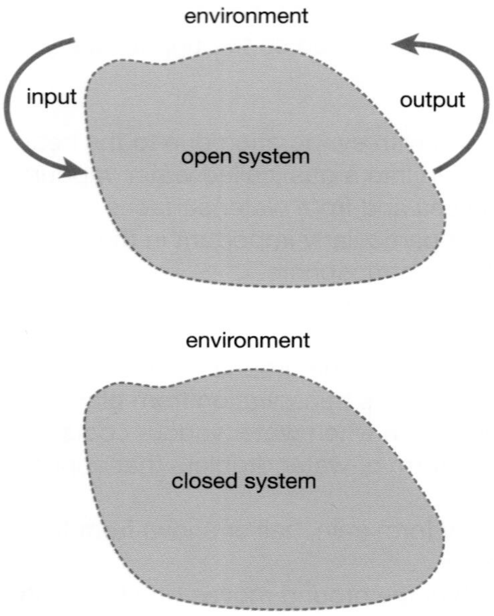

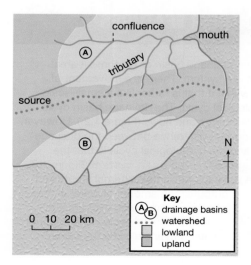

▲ Figure 1.3: Open and closed systems

CHECK YOUR UNDERSTANDING

Why is the hydrological cycle a closed system and the drainage basin cycle an open system?

CHECK YOUR UNDERSTANDING

Is there an inland sea in the country where you live?

▲ Figure 1.4: Basic features of the drainage basin

The inputs of a drainage basin are:

- energy from the Sun
- precipitation formed from moisture picked up outside the basin
- possibly water from tributary drainage basins – this is not shown in Figure 1.2; an explanation is given below.

The outputs are:

- the river's **discharge**
- the water in its basin from which evaporation and transpiration take place; this water eventually falls as precipitation in another drainage basin.

As Figure 1.4 shows, we can draw a dividing line between neighbouring drainage basins. It follows the tops of the hills and is called the **watershed**. The main river has its source in the higher parts of the basin close to the watershed. This is where most precipitation falls. Smaller streams, or tributaries, enter the main river channel at locations known as confluences. The mouth or **estuary** of the river is where it flows out into the sea.

Drainage basins can be of at least three broad types as follows.

- Those that simply collect and deliver water directly to the sea.
- Those that are parts of much larger drainage basins. For example, the basin of the River Negro in Brazil is a tributary or part of the huge drainage basin of the River Amazon. The water it collects is conveyed to the Amazon and then eventually to the sea.
- Some drainage basins do not lead, either directly or indirectly, to the open sea. Rather they lead to 'inland' seas or lakes such as the Caspian Sea located between Europe and Asia, the Aral Sea to the east of it, or Lake Victoria in East Africa.

The **channel** (or drainage) **network** is the system of surface and underground channels that collects and transports the precipitation falling on the drainage basin.

Figure 1.4 shows the channel networks of two drainage basins, A and B. Basin A's network is slightly less dense (has fewer channels) than B's, though the main channel of B is shorter. Channel networks can be mapped and their lengths and densities (number of channels per unit area) measured. The network can change over short periods of time. For example, during flooding drainage basins often have many more and longer channels than they do in periods of low or normal rainfall.

Each drainage basin is unique in its combination of features. These features include size, shape, rock type, relief and land use. They determine how quickly or slowly water moves through the basin.

Figure 1.5 shows how some of these features can affect overland flow or run off. Rock type and relief are physical factors over which people have little control. But land use can be easily changed by people. Woodland holds water and slows overland flow. However, once it is cleared for cultivation, run off will speed up. The built-up areas of towns and cities can speed up run off even more. Rainwater hits solid surfaces such as roofs, pavements and roads. It is then quickly channelled into drains which speed its delivery into a stream or river.

▶ **Figure 1.5: Factors affecting run off**

ACTIVITY

Suggest **two** reasons why drainage basins close to one another might have different drainage densities.

CHECK YOUR UNDERSTANDING

Be sure you know the difference between permeable and impermeable rocks.

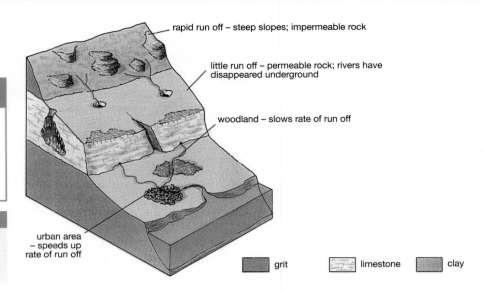

1.3 RIVER REGIMES AND HYDROGRAPHS

We need to know how quickly any rain falling in a drainage basin will reach the drainage network. It is also important to know how much a river's channel can hold. If rainwater reaches the river quickly, the channel may not cope and flooding will occur. The amount of water carried by a river at any one time is known as its discharge. This is measured in cumecs – that is, in cubic metres of water per second moving past a particular point along the river's course.

▶ **Figure 1.6: A hydrograph of the River Ganges in Bangladesh. Average monthly discharge is shown in cubic metres per second.**

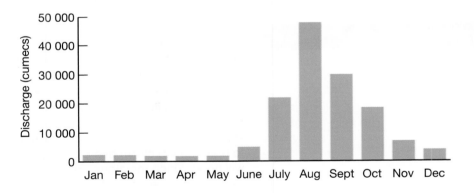

RIVER DISCHARGES

River discharges vary throughout the year, from month to month, from day to day. These variations make up what is termed the **river regime**. In most rivers, the regime closely reflects local climatic conditions, particularly the rainfall. Figure 1.6 shows the average monthly discharge of the River Ganges as it passes through Bangladesh. This diagram is a **hydrograph**. Clearly, mean (average) discharge is high between June and October. This period of high discharge coincides with the monsoon season, during which total rainfall can exceed 2750 mm.

▶ **Figure 1.7: Hydrograph of the River Thames at Reading, 2001–02**

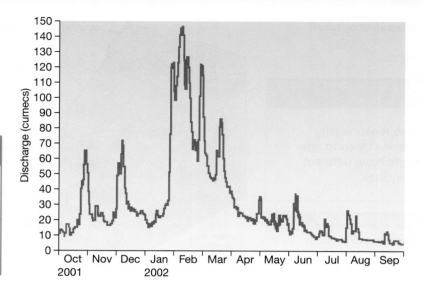

ACTIVITY

Compare the hydrograph of the Ganges with that of the Thames as shown by Figures 1.6 and 1.7. Try to explain the differences.

Figure 1.7 shows the regime of the River Thames (England) during one year. Unlike the Ganges, the highest discharges generally occur in winter, in February and March. Another feature of the hydrograph is its 'jaggedness'. Discharge varies from day to day. The peaks indicate the impact of passing showers and short periods of heavy rainfall.

The majority of the world's drainage basins are home to many people. The big attractions of such areas are their fertile soils and the ability to grow food. Today much money is invested in drainage basins, not just in farmland, but also in homes, businesses and transport. It is therefore important to know how rivers will behave following heavy rainfall. This data allows people to work out the risk of flooding, and over what area. This is where a storm hydrograph is useful.

STORM HYDROGRAPHS

A storm hydrograph records the changing discharge of a river after a rainstorm. The bars in the left-hand corner of Figure 1.8 show the input of rain. After rain hits the ground, it takes time for rainwater to reach the river and cause river levels to rise. This delay between peak rainfall and peak discharge is called the lag time. The shorter the lag time, the quicker the water reaches the river channel. A short lag time causes the river discharge to rise steeply. The steeper the rise in discharge, the greater the chances of flooding. It is possible to mark on the storm hydrograph the level of discharge above which the river will flood. Once the storm and its peak discharge have passed, the amount of water in the river starts to decrease.

▶ **Figure 1.8: A storm hydrograph**

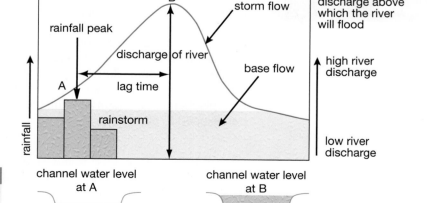

CHECK YOUR UNDERSTANDING

Explain why lag time is so important.

The storm hydrograph shows discharge of the river as being made up of two flows:

■ the **base flow** – the 'normal' discharge of the river
■ the **storm flow** – the additional discharge of the river as a result of the rainstorm.

FACTORS AFFECTING RIVER REGIMES

Six factors affect river regimes in general and storm hydrographs in particular.

■ The amount and the intensity of the rain. Heavy rain will not sink into the ground. Instead, it will become overland flow or run off and quickly reach the river.
■ Temperatures affect the form of precipitation. For example, if temperatures are below freezing, precipitation will be in the form of snow. This can take weeks to melt. If the ground remains frozen, melting snow on the surface can reach the river quickly.
■ Steep slopes will cause rapid surface run off, so water will reach the river more quickly. Flat and gently sloping land may lead to water sinking into the soil. This will delay it reaching the river.
■ Rock type – impermeable rocks will not allow rainwater to sink into it, so will speed up run off. Permeable rock allows infiltration and percolation of water into the bedrock. This in turn slows delivery of water to the river.
■ Vegetation and land use – trees and other plants intercept and delay the rain reaching the ground. Bare soil and rock speed up run off and reduce the time lag. So, too, will urban areas under tarmac and concrete.
■ Human intervention – **dams** and reservoirs are an obvious form of intervention in river regimes. By holding back discharge, dams reduce the risk of flooding downstream. The flow of water out of reservoirs can also be controlled by opening and closing sluice gates. The increasing **abstraction** of water from rivers for a range of growing human needs (see Part 1.7) impacts on the regimes of many rivers, not just on their storm hydrographs.

ACTIVITY

Draw an annotated diagram showing impacts of the six factors on river regimes.

1.4 FLUVIAL PROCESSES

WEATHERING AND MASS MOVEMENT

▲ **Figure 1.9: A landslide in Sri Lanka**

▶ **Table 1.1: Different types of weathering**

Rivers play a major part in shaping landforms. Three processes are at work here – **erosion**, **transport** and **deposition**. These river processes partner two other processes – **weathering** and **mass movement**. Let us first look at weathering and mass movement and then examine more closely the work that rivers do.

Weathering involves elements of the weather, particularly rainfall and temperatures (Table 1.1).

PHYSICAL WEATHERING	This breaks rocks down into smaller and smaller pieces. It is done by changes in temperatures and by rainfall freezing and thawing in rock cracks.
CHEMICAL WEATHERING	This causes rocks to decay and disintegrate. It is largely done by slightly acidic rain seeping into porous rocks.
BIOLOGICAL WEATHERING	The roots of plants, especially trees, growing into cracks in the rocks gradually split the rock apart.

All this destructive activity takes place where rocks are found above the surface of the surrounding land. Once rocks are really broken down, the weathered material starts to move down the slope under the influence of gravity. This is mass movement. It takes several forms. In river valleys, there are two main types of mass movement as follows.

- Slumping – this occurs when the bottom of a valley side slope is cut away by the river flowing at its base. It makes the slope unstable and weathered material slumps down towards the river. Slumping is also helped when the weathered material on the slope is saturated by heavy rain. The water does two things. It makes the weathered material heavier and acts as a lubricant. Figure 1.9 is an example of sudden slumping leading to a major landslide.
- Soil creep – weathered material moves slowly down slope under the influence of gravity. It collects at the bottom of the valley side and is eroded by the river.

EROSION

There are several different ways in which rivers erode their channels and valleys (Table 1.2).

HYDRAULIC ACTION	Water hits the river bed and banks with such force that material is dislodged and carried away. This is most likely to happen during periods when the river's discharge is high.
ABRASION	The material being carried by a river is rubbed against the sides and floor of the channel. This 'sandpaper' action widens and deepens the channel.
CORROSION (SOLUTION)	Minerals in the rocks forming the sides of the river channel are dissolved by the water flowing past them.

▲ Table 1.2: The processes of river erosion

Attrition is another river process. It involves particles of material being carried by a river, and becoming rounder and smaller as they collide with each other. This process does not cause erosion of river channels and valleys.

TRANSPORT

This is the movement of material (known as the **load**) by the river. The load is material that has been washed or fallen into the river. It also contains materials eroded by the river from the sides of the channel. The load can be transported in a number of different ways (Figure 1.10).

▶ Figure 1.10: Ways in which rivers transport their load

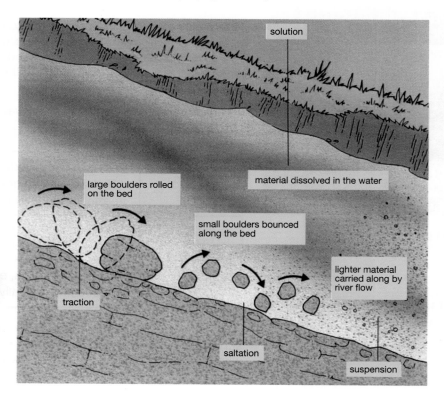

solution

material dissolved in the water

large boulders rolled on the bed

small boulders bounced along the bed

lighter material carried along by river flow

traction

saltation

suspension

DEPOSITION

Deposition is the laying down of material transported by the river. This occurs when there is a decrease in the energy, speed and discharge of the river. Deposition is most likely to happen when a river enters a lake or the sea. It also happens wherever there is a decrease in the gradient of the river's channel.

Erosion, transport and deposition are affected by a number of factors. A wetter climate means more discharge and therefore more erosion and transport. Softer rocks are more easily eroded and transported. Gentle slopes encourage deposition.

CHECK YOUR UNDERSTANDING
What sort of material is transported and deposited by rivers?

1.5 DOWNSTREAM CHANGES IN RIVER CHARACTERISTICS

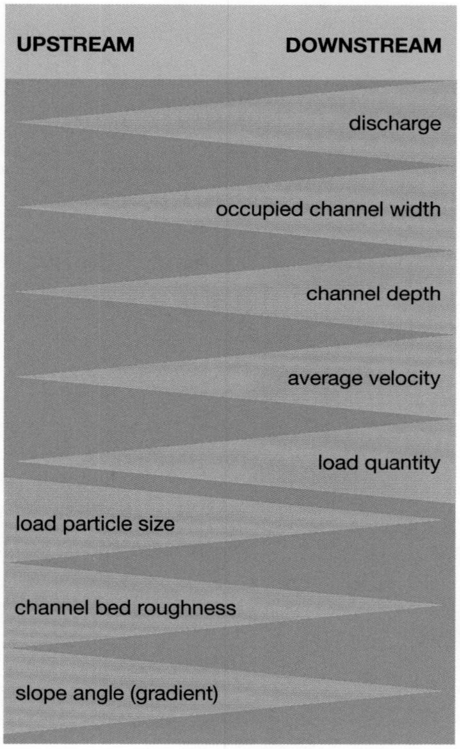

The **long profile** of a river runs from its source to the point where it enters the sea, a lake or joins another and larger river. The character of the long profile changes downstream. Overall, it has a smooth concave shape. It is steep and in places irregular where the river is flowing well above sea level in upland country. The irregularities occur where outcrops of hard rock run across the valley. Natural lakes and reservoirs can disrupt the smoothness of the long profile. However, the profile becomes much gentler and smoother as the river runs through lowland country and reaches its destination.

Of the many changes that take place along the long profile, changes to the river channel are particularly striking. The channel starts narrow and shallow (it is V-shaped) with rough edges. It gradually becomes wider, deeper and smoother. Figure 1.11 shows also that discharge and average velocity increase downstream, as does the amount of load being transported. At the same time, the size of sediment in the load becomes progressively finer and more rounded.

SKILLS REASONING

▲ Figure 1.11: Long-profile changes in channel characteristics

ACTIVITY

Explain the downstream changes, as shown in Figure 1.11, to each of the following:

- load particle size
- load quantity
- channel bed roughness
- channel depth.

1.6 DOWNSTREAM CHANGES IN RIVER LANDSCAPES

The downstream changes in channel characteristics are associated with changes in river landforms and landscapes. This can best be understood by dividing the long profile and its landforms into two sections: upland and lowland.

UPLAND LANDFORMS

CHECK YOUR UNDERSTANDING

In addition to slope steepness, what else would encourage mass movement on the valley side?

The main river landforms in upland areas are: steep V-shaped valleys, **interlocking spurs**, **waterfalls** and **gorges** (Figure 1.12). They have all been formed mainly by the processes of river erosion already described in Part 1.4. The processes of weathering and mass movement have also played a part.

In the uplands, the long profile is steep and the river flows fast. Much of the river's energy is spent in cutting downwards. The processes of hydraulic action (erosive force exerted by water) and abrasion, in particular, erode the river bed and make the valley deeper (Figure 1.12A). Because of the steepness and deepness of the valley, there is mass movement of material down the sides of the valley. Some of this material becomes river load and helps the abrasion process. The valley floor is narrow and often completely occupied by the river.

Interlocking spurs are formed where the river swings from side to side (Figure 1.12B). Again, the main work of the river is cutting vertically downwards into its bed. This means that the river cuts down to flow between spurs of higher land on alternate sides of the valley.

▶ **Figure 1.12: The origin of river landforms in upland areas**

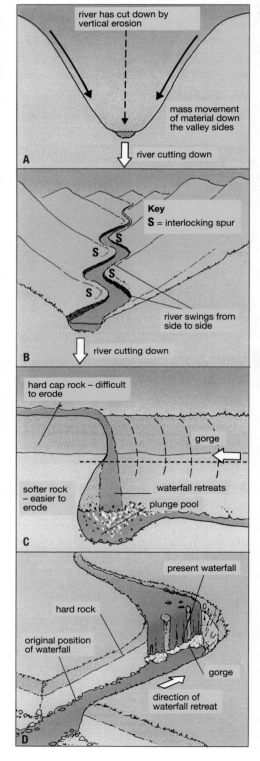

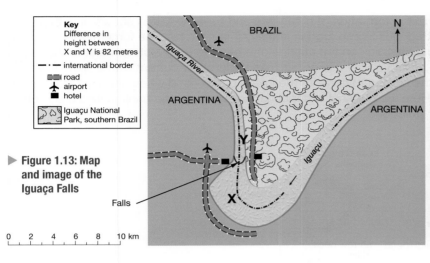

Key
Difference in height between X and Y is 82 metres
– · – international border
▨ road
✈ airport
■ hotel
Iguaçu National Park, southern Brazil

▶ Figure 1.13: Map and image of the Iguaça Falls

Falls

0 2 4 6 8 10 km

SKILLS   SELF DIRECTION

ACTIVITY

Can you name a waterfall located in your country?

Waterfalls occur where a band of hard rock is much more resistant to erosion than the softer rock below it (Figures 1.12C and D). This softer rock is readily eroded by the force of the water as it falls over the hard cap rock. Gradually, the falling water excavates a plunge pool at the bottom of the falls. Slowly, the hard rock is eroded back by the river and so the waterfall gradually retreats upstream leaving a gorge below it. The gorge is protected from being widened by its capping of hard rock. The Iguaça Falls in South America are a spectacular example and attract many tourists (Figure 1.13).

LOWLAND LANDFORMS

The river and its landforms change when the river leaves the uplands and flows across lowlands. The river channel and its valley become wider, deeper and smoother. Because of this, both the **river velocity** (speed) and the discharge of the river continue to increase, despite the gentler gradient. The river course in plan (map) view becomes less straight.

The valley cross-section is wider and flatter. The floor is occupied by a **flood plain**. Near the end of its course, the flood plain spreads out to become either a **delta** or an estuary.

ACTIVITY

Describe how river erosion changes downstream.

During its lowland course, the river is still actively eroding. However, vertical erosion is less important because the river is too close to sea level. More important is lateral erosion where the river wears away the sides of the channel, especially on the outside of **meanders**. The river becomes an agent of deposition as well. Such a large load of material has been picked up that, once the river loses energy, it drops some of that load on the flood plain. This is usually mud, stones and other organic matter. Every time the river leaves its channel, its velocity decreases. Once this happens, sediment is deposited across the valley floor. A great thickness of sediment builds up. The largest amount of deposition is always on the banks of the channel, which builds up to a greater height than the rest of the flood plain to form **levees** (Figure 1.14).

▼ Figure 1.14: The formation of a flood plain and levees

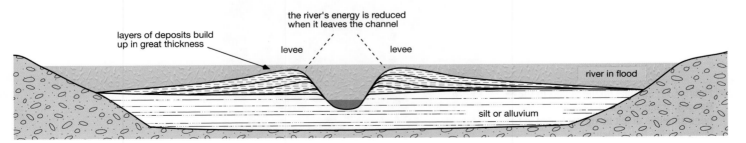

the river's energy is reduced when it leaves the channel

layers of deposits build up in great thickness

levee   levee

river in flood

silt or alluvium

A study of the formation of meanders and **ox-bow lakes** shows how the river both deposits and erodes laterally (Figure 1.15). The force of the water undercuts the bank on the outside of a bend to form a steep bank to the channel, called a river cliff or bluff. An underwater current with a spiral flow carries the eroded material to the inside of the bend where the flow of water is slower. Here the material is deposited to form a gentle bank, called a slip-off slope or point bar.

▼ Figure 1.15: Formation of meanders and ox-bows

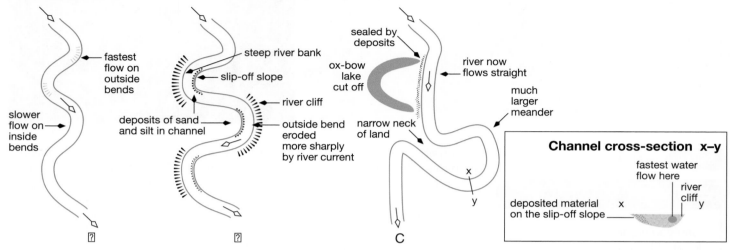

As lateral erosion continues, the bend of a meander becomes even more pronounced (Figure 1.15). Especially in times of flood, when the river's energy is much greater, the narrow neck of the meander may be breached (broken and crossed) so that the river flows straight again. The redundant meander loop keeps some water and becomes an ox-bow lake. Deposition during flooding seals off the edges and ends of the lake. Figure 1.16 shows a small part of the complicated meander pattern of the Mississippi River (USA).

SKILLS ▶ SELF DIRECTION

ACTIVITY

Find out this information about the Mississippi River:

■ the actual length of the river from its source to the sea
■ using a map, work out the straight-line distance from its source to the sea.

What do these **two** pieces of information tell you about the amount of meandering the river does?

▲ Figure 1.16: Mississippi meanders and ox-bows

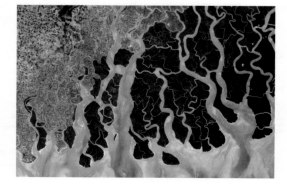

▶ Figure 1.17: Satellite image of part of the Ganges-Brahmaputra delta. It also shows huge amounts of silt being deposited in the waters of the Bay of Bengal.

ACTIVITY

Describe the physical features of the delta shown in Figure 1.17.

DID YOU KNOW?

Not all rivers build up deltas as they enter the sea. Many have open mouths or estuaries. Can you name **two** such examples, one in your own country and one elsewhere?

The delta is the final landform of the river's journey from its source to the sea or lake. Deltas are vast areas of alluvium at the mouths of rivers. The Ganges–Brahmaputra delta is one of the largest in the world (Figure 1.17). It is located at the head of the Bay of Bengal. The rivers that have built it up carry huge quantities of sediment, mainly from the Himalayas, down to the delta. It is estimated that they deliver about 1.7 billion tonnes of sediment each year. The flows of the combined rivers are slowed as they meet the denser sea water in the bay. The result is that much of the load is dropped. In fact, the load is deposited faster than the tides can remove it out to sea. The river flow is blocked by deposition so that the rivers split up into smaller channels known as distributaries. These distributaries deposit sediment over a wide area, creating new land where there was once sea.

CASE STUDY: THE RIVER TAY (UK) AND ITS VALLEY

IN THE UPLANDS

The River Tay is fed by streams which drain the slopes of the Grampian Mountains in the Highlands of Scotland. Precipitation in the upland parts of the drainage basin is high (well over 1000 mm a year) and slopes are steep.

ACTIVITY

Look at an atlas of the British Isles to locate the drainage basin of the River Tay.

▶ Figure 1.18: Part of the upland course of the River Tay

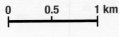

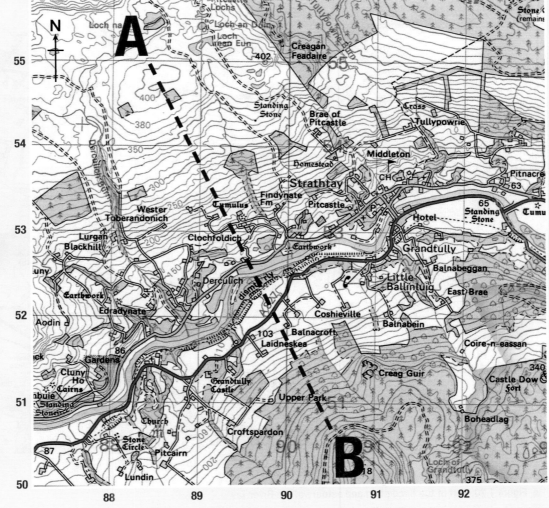

The height of the drainage basin and the steep slopes result in large amounts of run off. The Tay is already about 100 metres across in that part of its course, as shown on the Ordnance Survey map extract (Figure 1.18). The curving river course suggests that it is flowing between interlocking spurs.

The valley cross-section is shown in Figure 1.19A. It is V-shaped and steep-sided. The river fills the valley floor. The cross-section shows that the river is still flowing at some height above sea level. Erosion by the river appears to be vertical rather than lateral.

IN THE LOWLANDS

Figure 1.19B shows the valley cross-section near the sea. The flat and low-lying land is the flood plain. It is 600 m wide where the tributary River Earn meets the main River Tay. Notice the big meander loop on the tributary through grid squares 1718 and 1717 (Figure 1.20). The black dashes marked around its banks show the levees.

▼ Figure 1.19: Cross-sections across the Tay valley

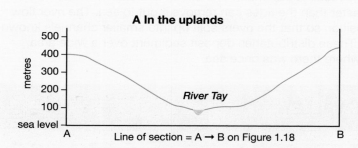

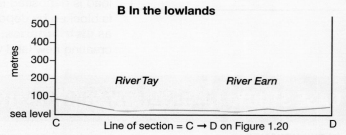

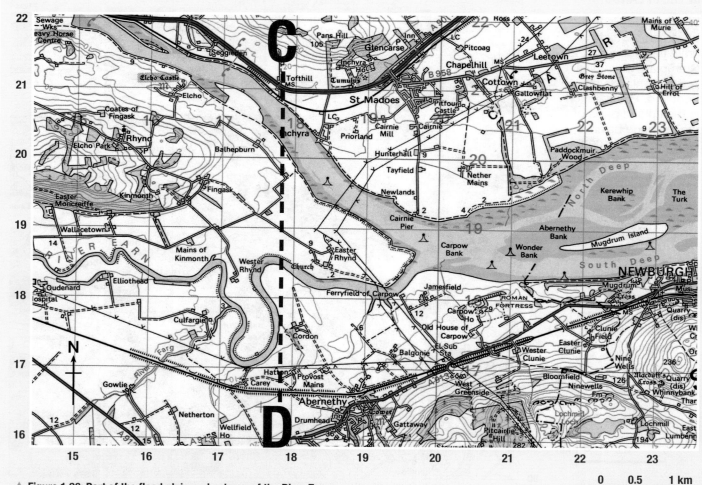

▲ Figure 1.20: Part of the flood plain and estuary of the River Tay

The River Tay shows many of the typical estuary features that can be found at the mouths of rivers (Figure 1.20). They include:

- a wide channel – up to 2 km
- sand and mud banks (such as Abernethy Bank)
- some areas of marsh (in grid square 2119)
- channels of deeper water (such as North Deep).

SKILLS EXECUTIVE FUNCTION

ACTIVITY

Using Figure 1.20, draw a sketch map of the River Tay estuary to the east of easting 20.

SKILLS INTERPRETATION

ACTIVITY

What are the advantages and disadvantages of the Tay estuary for shipping?

1.7 WATER USES, DEMAND AND SUPPLY

USES

Only 3 per cent of all water on Earth is freshwater; the rest is salt water in the seas (Figure 1.21). Over 75 per cent of the fresh water is locked up in glaciers and ice sheets, and 20 per cent is groundwater. The world's remarkably small amount of fresh water is:

- essential to all life
- vital to economic development
- unevenly distributed, with some areas 'water-rich' and others 'water-poor'.

SKILLS DECISION MAKING

ACTIVITY

Discuss, in groups, how global warming might change Figure 1.21. Compare the conclusions from different groups.

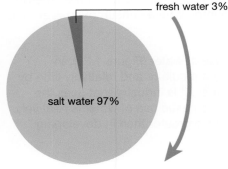

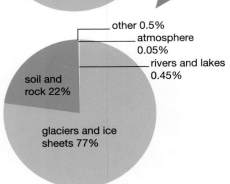

Fresh water is needed for:

- domestic use – bathing and showering; flushing toilets; drinking and cooking; washing clothes and dishes; watering the garden
- industrial use – producing a wide range of goods from beer to steel; generating electricity
- agricultural use – irrigating crops; providing drinking water for livestock
- leisure use – sport fishing on rivers; sailing on lakes and ponds; watering golf courses.

All forms of water use revolve around two key elements:

- demand – this is the need for water for a range of uses; it is also often referred to as **consumption**, and the amount of water consumed reflects the level of demand
- supply – meeting the demand for water by tapping various sources, such as groundwater, lakes and rivers.

▲ Figure 1.21: The world's water

CHECK YOUR UNDERSTANDING
Check that you know: a) the main sources, and b) the main uses of fresh water.

For any country or area within it, we can compare water demand and water supply. This comparison is known as the **water balance** (Figure 1.23 on page 19).

DEMAND

Water demand and consumption have greatly increased during the last 100 years with global demand more than trebling over the last 50 years. The major factor behind this rising demand has been the continuing growth of the world's population. Between 1990 and 2015, the global population grew from 5.3 billion to an estimated 7.4 billion. However, the rise in the demand and consumption of water is not just due to more people. Aspects of development also play an important part.

SKILLS STATISTICAL

ACTIVITY

Find out more about water supply in your home country. By how much has the demand for water increased in recent years? Who are the main water users?

- The rising standard of living that is part of development increases the domestic use of water. More houses with piped water, flush toilets, showers and baths, washing machines, and even swimming pools mean much higher water consumption.
- The rise in agricultural productivity needed to feed a growing population increases the use of water, particularly for **irrigation**.
- Industrialisation is a key part of development. Most factories are large consumers of water. Water is used for cooling machinery. It is also used to generate electricity for powering industry.

In many parts of the world where water is in short supply and obtained mainly from wells, it is women who take water from the well to the home. It is hard work, often over long distances. Many people believe that the expectation that women will do this work symbolises a lack of gender equality.

Given the link between development and water consumption, we can recognise two 'worlds of water'. There are big differences between water consumption in developed and developing countries. Water consumption in developed countries is very high. On average, each person in these countries uses about 1200 cubic metres of water each year. This is about three times as much as a person living in a developing country, where consumption is around 400 cubic metres per year.

There are also some big differences in the use of water (Figure 1.22). In developing countries, most water is used by agriculture and relatively little by industry or in the home. In developed countries, it is industry that uses the most water followed closely by agriculture. Domestic use of water is relatively small, but as a percentage it is over three times greater than in developing countries.

▶ **Figure 1.22: Water consumption in developing and developed countries, 2015**

SKILLS DECISION MAKING

ACTIVITY

Discuss, in class, ways in which climate may affect the demand for water.

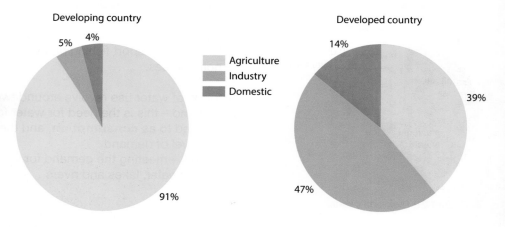

Developing country

5% 4%

91%

Developed country

14%

39%

47%

Agriculture
Industry
Domestic

SUPPLY

SKILLS SELF DIRECTION

ACTIVITY

Find out where most of the water supply in your home country comes from.

WATER SURPLUS AND DEFICIT

Having looked at one side of the water equation, let us now look at the other – water supply. In many parts of the world, the water needed to meet increasing demand comes from three main sources as follows.

- Rivers and lakes – possibly the source used by the earliest humans.
- Reservoirs – artificial lakes created by building a dam across a valley and allowing it to flood. The water collected and stored behind the dam can become an important water supply.
- Aquifers and wells – much of the world's fresh water supply lies underground. It is stored in porous rocks known as aquifers. This groundwater can be extracted by drilling wells or boreholes down to the aquifer. The water is then raised to the surface by buckets, pumps or under its own pressure.

If we made two world maps, one showing distribution of water demand or consumption, and the other showing water supply, we would find that these maps show quite different patterns. If we laid one map on top of the other, we would be able to pick out three types of area:

- areas where the water balance is negative – where water demand exceeds supply; these are referred to as water-deficit areas
- areas where the water balance is positive – these are water-surplus areas where the supply or availability of water exceeds demand
- areas where water demand and supply are roughly the same – water-neutral areas where the water balance is equal.

Figure 1.23 shows the results of overlaying two world maps – one of demand and the other of supply. Once again, we can see 'two worlds of water'. There are very few truly water-surplus areas. However, in large areas the water balance is slightly in surplus. These are typically remote, mountainous regions with high annual rainfall, few people and low water demand. However, there are many water-deficit areas. These are most obvious in Africa and the Middle East, in Australia and parts of North and South America. Many of these areas are in deficit because they receive little precipitation during the course of a year. Others are in deficit because of large populations and rising development. A good example of such an area is India.

There are various ways of moving water from surplus to deficit areas. For example, by hand in plastic bottles and buckets, by motor vehicles and even by tanker ships. However, the most widely used way is by long-distance pipelines and canals (see Case Studies in Part 1.8).

▶ Figure 1.23: Water-surplus and water-deficit areas of the world

SKILLS INTERPRETATION

ACTIVITY

The shades of green in Figure 1.23 show those parts of the world where water supply just about meets water demand. Describe the main areas of the world where this is the case.

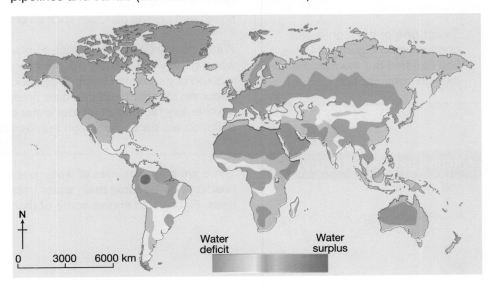

Water deficit   Water surplus

0   3000   6000 km

1.8 WATER QUALITY AND SUPPLY

In Part 1.7, we looked at water demand and supply simply in terms of quantities. Quantity is important. Figure 1.24 shows how much water is used to provide things that we use in everyday life.

▼ Figure 1.24: The amount of water it takes...

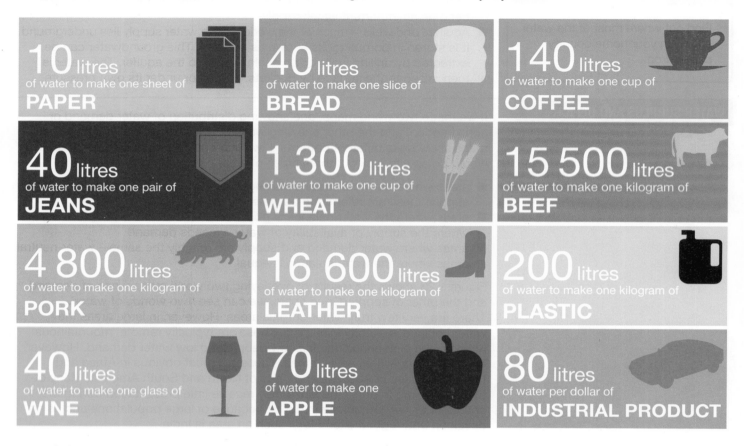

10 litres of water to make one sheet of **PAPER**

40 litres of water to make one slice of **BREAD**

140 litres of water to make one cup of **COFFEE**

40 litres of water to make one pair of **JEANS**

1 300 litres of water to make one cup of **WHEAT**

15 500 litres of water to make one kilogram of **BEEF**

4 800 litres of water to make one kilogram of **PORK**

16 600 litres of water to make one kilogram of **LEATHER**

200 litres of water to make one kilogram of **PLASTIC**

40 litres of water to make one glass of **WINE**

70 litres of water to make one **APPLE**

80 litres of water per dollar of **INDUSTRIAL PRODUCT**

SKILLS INTERPRETATION

ACTIVITY

Select **one** of the high-water consumers in Figure 1.24; see if you can find out the reasons for its high consumption.

However, in producing most of the items shown in Figure 1.24, water quality is also important. This is obvious where water is used for domestic purposes such as drinking, cooking and washing. But clean water is also needed for growing crops and rearing livestock. Clean water is a priority. Polluted water threatens human health and causes diseases such as cholera, bilharzia, typhoid and diarrhoea.

Water quality varies from place to place for a variety of reasons. For example, water quality is generally poorer in dry climates or where the climate has a marked dry season. During dry periods, any water that remains on the surface becomes stagnant and can be a breeding ground for diseases. **Pollution** is another key factor that affects water quality, and this varies greatly. Levels of pollution are particularly high in urban areas, especially in developing cities.

SOURCES OF WATER POLLUTION

There are many sources of water pollution. We can group them under the headings of the three main water users – agriculture, industry and domestic uses. Figure 1.25 shows some of them in the 'cycle of water pollution'.

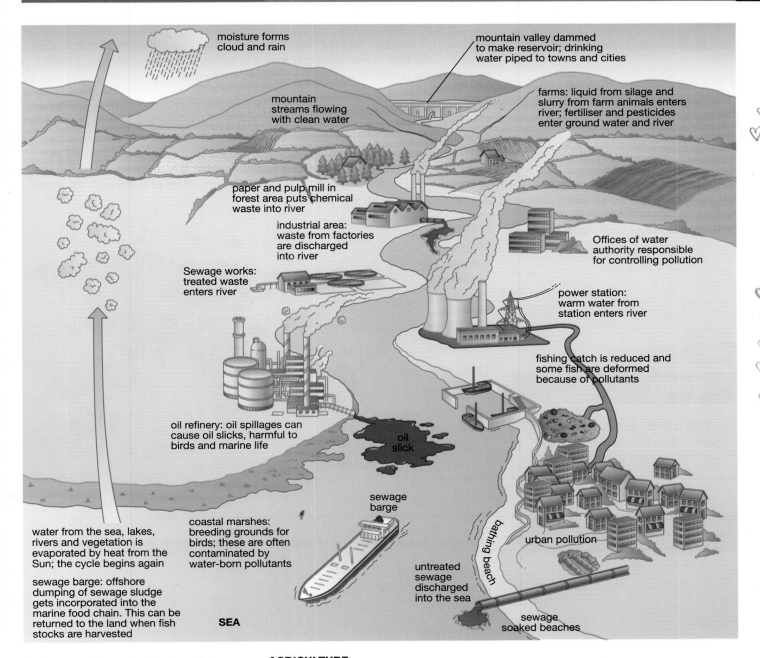

▲ **Figure 1.25: The cycle of water pollution**

SKILLS CRITICAL THINKING

ACTIVITY

Can you think of ways in which recreation and tourism pollute water?

AGRICULTURE
- Liquid from farm silage and slurry from farm animals enters rivers.
- Fertilisers and pesticides seep into the groundwater.
- **Deforestation** – run off carries soil and silt into rivers, with serious effects on aquatic life and humans who drink the water.

INDUSTRY
- Taking cooling water for an electric power station from a river and returning it to the source at a higher temperature upsets river ecosystems.
- Spillages from industrial plants such as oil refineries can enter rivers.
- Working of metallic minerals and the heavy use of water in processing ore – toxic substances from this eventually find their way into rivers.

DOMESTIC
- The discharge of untreated sewage from houses – even treated sewage pollutes.
- Use of river for washing clothes and bathing contaminates water.
- Emptying highly chlorinated water from swimming pools contaminates water.

Safe water is water that is fit for human consumption. It is not contaminated by pollutants and is free from disease. It is estimated that more than 1 billion people in the world do not have access to safe water.

Figure 1.26 shows that countries with the lowest access to clean water (where less than 50 per cent of the population enjoy safe water) are mainly located in Africa and in parts of southern Asia. As a consequence, many people suffer ill health and an early death. In most developed countries, more than 90 per cent of the population have access to clean water. It is surprising that the map shows there is no data about clean water for a number of European countries. It is difficult to find an explanation for this.

▶ Figure 1.26: Access to safe water

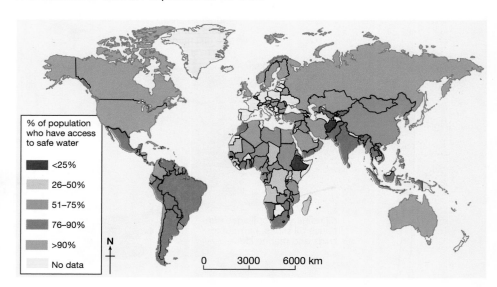

SKILLS INTERPRETATION

ACTIVITY

Write a short account describing the main features shown by Figure 1.26.

MANAGING THE SUPPLY OF CLEAN WATER

Managing the supply of clean water involves three main stages (Figure 1.27):

■ collection
■ treatment
■ delivery.

COLLECTION
In Part 1.7 the main sources of water were listed as:

■ rivers
■ reservoirs and lakes
■ aquifers and wells.

TREATMENT
It is very rare to use water that does not need some form of purification. Rivers are often highly polluted. Reservoir water can be polluted by acid rain and the seepage of pollutants from surrounding hillsides. It used to be thought that groundwater was pure. However, groundwater can be badly contaminated by chemicals in the rocks. In parts of Bangladesh, groundwater is contaminated by arsenic, a highly toxic chemical.

Water treatment aims to remove pollutants from the 'raw' collected water and produce water that is pure enough for human consumption (Figure 1.27). Substances that are removed during the process of treating drinking water

▲ Figure 1.27: Three stages in the supply of clean water – a) collection (reservoir), b) treatment (waterworks) and c) delivery (standpipe)

include suspended solids (silt and soil), bacteria, algae, viruses, fungi, minerals (iron, manganese and sulphur), and synthetic chemical pollutants (fertilisers).

A combination of processes is used worldwide in the treatment of water for human consumption. The combination varies according to the nature of the raw water. The processes include:

- chlorination – to control any biological growth (e.g. algae)
- aeration – to remove dissolved iron and manganese
- sedimentation – to remove suspended solids
- filtration – to remove very fine sediments
- disinfection – to kill bacteria.

DELIVERY

In many countries, particularly developed countries, water is delivered from the treatment works to the point of consumption (the home, factory, etc.) by pipes. The costs of installing and maintaining pipe networks for the distribution of water are high. Unless the network is properly maintained, there can be large losses of water from leaking pipes.

In the urban areas of many developing countries, water is often delivered by standpipes (pipes through which water is pumped vertically) in the streets. In villages, wells are important water sources. Often, well water is used untreated. Where water is collected in buckets and carried from a well or standpipe into the home, there are further risks of pollution. Buckets and containers may be dirty. Water left in open buckets in the home is vulnerable to pollution by dust, insects and animals. A universal problem with the delivery of water by standpipes and wells is the time spent collecting it.

There is a fourth and relatively new way of delivering water – in plastic bottles filled at a source such as a spring. This is a very expensive way to deliver water because of the cost of the plastic containers, and of filling and transporting them to the point of sale. The biggest sales of bottled water actually occur in developed countries. Some people are worried that drinking piped domestic water might be risky. They are concerned about water quality and the efficiency of water treatment, even in developed countries.

Water management is not just about ensuring a good supply of clean water. Dams and reservoirs serve a number of different purposes, such as generating electricity, supplying irrigation water, exercising flood control and improving navigation. Reservoirs and lock systems allow boats and ships to travel much further upstream than previously. This multi-purpose use of water, or river management, is well illustrated by projects in Spain and China. The aim behind all these projects has been to increase the availability of water in areas of high water demand.

SKILLS ▶ CRITICAL THINKING

ACTIVITY

What do you think are the advantages and disadvantages of bottled water?

CASE STUDY: RIVER MANAGEMENT LESSONS FROM SPAIN

A Mediterranean climate means that Spain's annual rainfall is small. Most of it falls in the winter half of the year, so that summer droughts are common. Also, the availability of water in Spain and the demand for water do not coincide. The basic idea behind the two projects described below was to transfer water from the sparsely populated centre of Spain to provide much-needed water in the south-east, particularly for irrigation.

THE TAGUS-SEGURA PROJECT

A 286 km-long series of aqueducts and canals was completed in 1978 to carry water from two reservoirs in the upper Tagus valley to a large reservoir in the Segura valley (Figure 1.28). Water is distributed from this reservoir to consumers in the provinces of Alicante, Murcia and Almeria.

There are a number of problems as a result of this project:

- most of the irrigation water has gone to agribusinesses rather than local farmers
- much of the water has been consumed, not by agriculture, but by the urban developments and tourist resorts along the coast (Figure 1.29)
- the availability of this transferred water has encouraged a wasteful or extravagant use of it.

THE EBRO PROJECT

In 2001, an even more ambitious river management plan was proposed. This was to dam and divert the longest river in Spain, the Ebro, which flows into the Mediterranean some way south of Barcelona. The scheme meant building more than a hundred dams in the upper reaches of the Ebro as well as hundreds of kilometres of canals. These would transfer 100 billion litres of water per year to the same arid south-eastern parts of Spain as the water from the Tagus.

In 2004, the project was abandoned largely because of:

- the bad experiences of the Tagus-Segura project
- the huge threat it posed to the Ebro delta, an important agricultural area and wildlife wetland (Figure 1.30).

The Spanish government decided to meet the growing water shortages by constructing coastal desalinisation plants. This water policy continues to today.

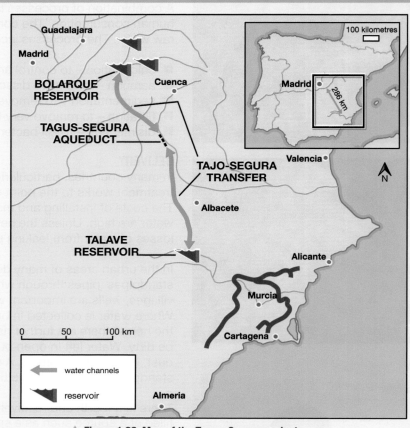

▲ Figure 1.28: Map of the Tagus-Segura project

▲ Figure 1.29: Cartagena is one of the consumers of Tagus water.

▲ Figure 1.30: An aerial view of the Ebro delta

SKILLS   SELF DIRECTION

ACTIVITY

Find out why it was so important to increase the supply of irrigation water to south-east Spain.

CASE STUDY: RIVER MANAGEMENT LESSONS FROM CHINA

Over the last 25 years, China has undertaken two major river management projects for slightly different reasons.

THE THREE GORGES PROJECT
This was started in 1997, mainly to generate electricity and to control flooding. Figure 1.31 shows the location of the project in the middle of China and on its longest river, the Yangtze. The dam was finished in 2009 and is still the largest in the world. It is 185 m high and almost 2 km wide. The reservoir of water that has built up behind the dam is nearly 600 km long. It has greatly improved water transport along this section of the Yangtze valley.

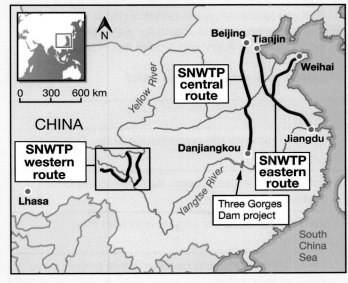

▲ Figure 1.31: The locations of two Chinese projects: Three Gorges Dam and the SNWTP (see the next Case Study)

▲ Figure 1.32: The completed Three Gorges Dam

Hydroelectric generators at the dam (Figure 1.32) produce about 10 per cent of China's present demand for electricity. But this demand continues to rise with the general development of the country and increases in population. In terms of the environment and global warming, it is important that China does not produce electricity by burning coal. Another advantage of the reservoir is that it will improve water transport along its length.

The project has, however, a downside. One million people lost their homes and had to be resettled to make way for the dam and its reservoir. Cities, towns and villages have been drowned. The reservoir has also flooded some of the country's most fertile land. There are longer-term costs. For example, silt is trapped behind the dam, making farmland lower down the Yangtze less fertile over time. Water quality is reducing as huge amounts of human and industrial waste are also trapped behind the dam. Another concern is whether the dam will

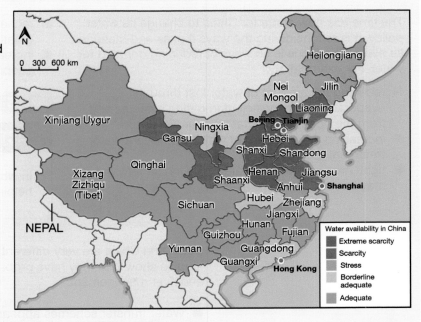

▲ Figure 1.33: Water availability in China

be able to cope with the frequent earthquakes that affect this part of China.

THE SOUTH-NORTH WATER TRANSPORT PROJECT (SNWTP)

Much of China's amazing economic and population growth is concentrated on the North China Plain, particularly around the megacities of Beijing and Tianjin. In this area, there is a huge demand for water. This demand is created by over 200 million people, the concentration of heavy industry and the irrigation of vast areas of farmland.

The SNWTP was created, as with the Tagus-Segura Project in Spain, because water demand was significantly greater than water availability (Figure 1.33). The idea of moving water from the 'surplus' South to the 'deficit' North was first discussed in 1952. Fifty years later, work started on the SNWTP to transfer 12 trillion gallons of water a year over a distance of 1000 km along three different routes (see Figure 1.33).

The eastern and central routes were opened in 2013 and 2014 respectively. These routes linked together existing canals, rivers and lakes. But there were human and environmental costs. Like the Three Gorges Dam project, both routes use the Yangtze River. Work on the western route has been halted largely because of:

- a growing awareness of those 'costs' which are now thought to outweigh the benefits (Table 1.3)
- a levelling off in the demand for water because population and economic growth has slowed down.

▼ Table 1.3: Costs and benefits of the SNWTP

COSTS	BENEFITS
The South is likely to suffer water shortages as it is exporting too much of its water to the North.	The North remains the powerhouse of the Chinese economy.
Large numbers of people have been displaced to make way for the transfer routes.	Much-needed water is being supplied to important industries.
Wildlife and ecosystems are badly disturbed by the water transfer routes.	There is more water for irrigation to help food production.
Loss of water as a result of evaporation from the open canals used to transfer the water.	Health risks are reduced because more people have access to safe water.
Very expensive project and a burden on taxpayers.	A showcase for Chinese engineering and technology.

The time has now come for China to change its water policy. It is reconsidering the ways it uses and manages its rivers. The focus is on ways of reducing demand for water. For example, by:

- reducing the amount of water lost through evaporation from inefficient irrigation systems and from the open canals that transfer water
- increasing the recycling of water by modern water treatment techniques
- tightening controls on water pollution so that water may be more easily re-used

- recharging groundwater stocks that have been greatly lowered by overuse
- increasing the price of water and so discouraging people and businesses from using it wastefully.

SKILLS   REASONING

ACTIVITY

In what ways are the two Chinese water projects helping the growth of its economy?

Spain and China are very different in size and level of development. The case studies show that they have probably learned the same lessons about water supply and river management.

- Water transfer schemes appear at first to be a most effective solution to the shortage of water in parts of the countries, but they do have serious financial, social and environmental costs.
- Both countries have moved away from ambitious water transfer projects, but in different directions. Spain now uses desalinisation to help meet water shortages, while China is searching for ways of using water more efficiently and sustainably.

CASE STUDY: PLANS TO MANAGE THE BLUE NILE, ETHIOPIA

A shortage of resources is one reason why Ethiopia is one of several countries experiencing the most developmental difficulty. Two resources in particular are in short supply – water and energy. These are shortly to be met in the basin of the upper Blue Nile. The headwaters of the Blue Nile start at Lake Tana, the largest lake in the country, in the Ethiopian Highlands. They are soon joined by many tributaries also draining the highlands. Where these tributaries come together they form a significant river before reaching the lowlands and crossing into Sudan. The distance between Lake Tana and the Sudan border is nearly 850 km with a fall of 1300 metres.

Work started on the Grand Renaissance Dam (Figure 1.34) in 2011 and is due to finish in 2017. It will generate 6000 megawatts of electricity and provide much-needed irrigation water. The project is being funded by the Ethiopian government. However, it has been controversial. This was mainly because Sudan and Egypt, both located downstream, felt that the dam would 'steal' too much of the Nile's water. But it is now recognised that the dam will also benefit these two countries.

The opening of this river management project is expected to start a new era in Ethiopia's development.

▲ Figure 1.34: The location of the Grand Renaissance Dam

ACTIVITY

Make a two-column table and list the advantages and disadvantages of the Grand Renaissance Dam project.

1.9 FLOODING – CAUSES AND CONTROL

Flooding occurs when the amount of water moving down a river exceeds the capacity of the river's channel. The excess water overflows the banks and spills out across the flood plain. Flooding is a hazard that can cause great damage to the environment and people.

CAUSES

Rivers usually flood as a result of a combination of causes. This combination can involve both natural (physical) and human factors. In many cases, natural flooding is made worse by people and their activities.

Most flooding is related to spells of very heavy rainfall. The critical factor is how quickly this rainfall reaches the river from where it falls on surrounding land and mountains. The shorter the lag time (see Figure 1.8 on page 8), the greater the chances of flooding.

▶ **Table 1.4: Factors causing flooding**

PHYSICAL FACTOR	IMPACT
Weather	Run off speeded up when prolonged heavy rainfall exceeds the infiltration capacity and the ground becomes saturated. Sudden rise in temperatures above freezing causes rapid snow melt.
Rock	Impermeable rocks limit percolation and encourage rapid surface run off or overland flow
Soil	A low infiltration capacity in some soils (e.g. clay) speeds run off.
Relief	Steep slopes cause rapid run off.
Drainage density	High drainage density means many tributary streams carry the rainwater quickly to the main river.
Vegetation	Low density vegetation absorbs little water and does not seriously slow down run off.
HUMAN FACTOR	
Deforestation	Cutting down trees reduces interception and speeds up run off.
Urbanisation	Concrete and tarmac surfaces, together with drains, mean quicker delivery of rainwater to the main river.
Agriculture	The risk of flooding is increased by leaving the soil bare, overgrazing and monoculture, and by ploughing down rather than across slopes.
Burning fossil fuels	This is raising global temperatures and causing more melting of ice sheets and glaciers, as well as more rainfall and more frequent storms.

▲ **Figure 1.35: Rescue services responding to flooding in Carlisle, 2015**

SKILLS   INTERPRETATION (NUMERICAL)

ACTIVITY

Which of the factors in Table 1.4 is likely to have the greatest effect on lag times? You may need to read the information on pages 8 and 9 about the storm hydrograph again.

Table 1.4 lists some of the factors that reduce lag times and therefore increase the chances of flooding.

Flooding can also result from persistent rain over a long period of time. Gradually, the water table (level of water underground) rises and the soil becomes saturated. Downward infiltration ceases and water simply accumulates in shallow depressions and on low-lying land. In December

2015, a quick succession of storms flooded many parts of northern England (Figure 1.35). The cost of dealing with the flood damage was around £2 billion. Unfortunately, too many businesses and householders found that they did not have adequate insurance.

CONSEQUENCES

▶ Figure 1.36: The effects of flooding on the environment and people

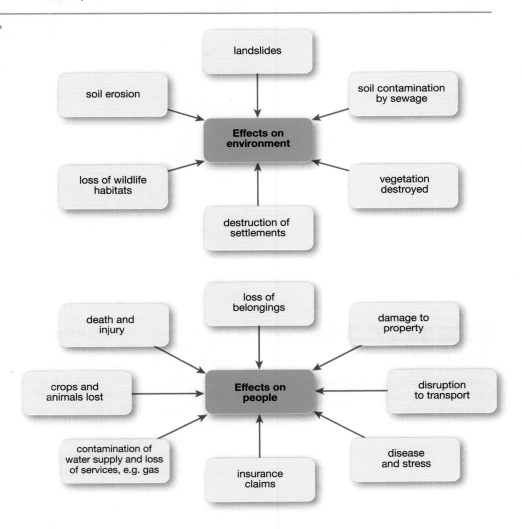

SKILLS CRITICAL THINKING

ACTIVITY

Suggest reasons why floods in developed countries are often more costly than those in developing countries.

Figure 1.36 shows some of the more common impacts of flooding. River floods can cause a lot of damage. As with all natural hazards, there are both immediate and long-term effects. The immediate effects include loss of life, destruction of property and crops, homelessness, the disruption of transport and communications, and the loss of water supply and sewage disposal services.

In the longer term, there is the cost of replacing what has been lost and damaged. One particular challenge is removing the huge amounts of silt deposited by the floodwaters as the waters go down. In developed countries, the risks may be covered by insurance.

Poor people in developing countries, however, may lose everything in floods. With cropland ruined and animals lost, widespread famine can result. Stagnant floodwater polluted by human excrement can become a serious health hazard. Under these conditions, diseases such as typhoid and cholera thrive. There is an urgent need for emergency food and health aid. The best hope of recovery lies in the help provided by international aid organisations.

CONTROL

▶ **Figure 1.37: Some forms of flood management**

Flood control and flood management can involve three different types of action (Figure 1.37).

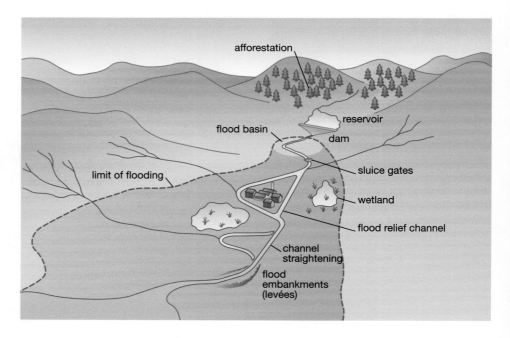

CHECK YOUR UNDERSTANDING

It is important that you are clear about:
a) the difference between hard engineering and soft engineering, and b) why soft engineering is more environmentally friendly.

- **Construction** – this is building **hard-engineering** structures such as dams, flood embankments (raised artificial banks), sluice gates and relief channels (spillways). These structures either hold back or help to safely dispose of floodwater. However, they are generally expensive to build.
- **Adjustment** (or **mitigation**) – flood control schemes that try to avoid or minimise potential flood damage are becoming popular. This involves working with nature rather than against it and is called the **soft-engineering** approach. For example, it might involve restoring a river to its natural state or preserving marshes and wetlands on flood plains to act as valuable temporary stores of floodwater (like sponges). Other actions include having stricter planning controls that minimise building on the flood plain. People in flood-risk areas can be encouraged to take out flood insurance. Better flood warning systems and publicising what to do in an emergency could also help.
- **Prediction** – prediction of river floods (their extent and depth) is important to flood control. Knowing how high or how wide a river can become during flood conditions helps people decide how high to build river embankments. It also helps to stop the building of houses, factories and services in areas where there is a high risk of flooding.

The problem with prediction is that floods vary. Small floods may occur fairly regularly, for example once every five years. Large floods happen less frequently – possibly once every 50 years. Perhaps every 1000 years there will be an exceptional flood on a huge scale. The question for river managers is to decide for which level of flood they should provide protection. This is known as risk assessment. Protecting against the regular five-year flood is relatively cheap and easy. However, it may not be worth investing huge sums of money to provide protection against the hugely damaging 'once-in-a-thousand-years flood'. In many cases, river managers will take a middle course and aim to deal with the medium threats of the 50- to 100-year floods. Clearly, if the risk is under-assessed, protection is likely to be inadequate.

SKILLS ▶ REASONING

ACTIVITY

Explain how improved weather forecasting might help the prediction of floods.

CHAPTER QUESTIONS

END OF CHAPTER CHECKOUT

INTEGRATED SKILLS
During your work on the content of this chapter, you are expected to have practised in class the following:

■ drawing and interpreting storm hydrographs
■ using geological maps to understand drainage characteristics
■ using GIS to map river systems
■ using different maps to investigate the impacts of human intervention
■ using and interpreting weather and climate data.

SHORT RESPONSE

1 Identify **one** input and **one** output of the drainage basin cycle. [2]

2 Suggest **two** reasons why flood plains are valuable to people. [2]

3 State how **one** named river channel characteristic changes downstream. [1]

4 Describe how a waterfall is formed. [3]

LONGER RESPONSE

5 Describe how a river transports its load. [4]

6 Examine the value of hydrographs. [6]

7 For a named country, analyse the sources of river pollution. [8]

8 Compare soft-engineering and hard-engineering approaches to the management of floods. [8]

EXAM-STYLE PRACTICE

1 Identify the meaning of the term 'watershed'. [1]
A The edge of a drainage basin
B The top of a waterfall
C The mouth of a river
D The soaking of water into the ground

2 a) Identify **one** store in the hydrological cycle. [1]
b) State **one** flow in the hydrological cycle. [1]
c) Explain **one** process of river erosion. [2]

3 Study Figure 1.24 (on page 20). Suggest **two** reasons why agriculture is such a large user of water. [4]

4 Explain the formation of an oxbow lake. [3]

5 Study Figure 1.7 (on page 8). Identify the month when the discharge of the Thames is greatest. [1]

6 Explain **two** ways in which people increase the flood risk. [4]

7 Study Figure 1.23 (on page 19). Analyse the factors affecting the distribution of high water deficits. [8]

[Exam-style practice total 25 marks]

2 COASTAL ENVIRONMENTS

LEARNING OBJECTIVES

By the end of this chapter, you should know:

- The physical processes that affect the coast

- The influence of geology, vegetation and sea-level changes on coastal environments

- The role of erosional and depositional processes in the development of coastal landforms

- The distributions and features of the world's coastal ecosystems

- The abiotic and biotic characteristics of one named coastal ecosystem

- How coastal ecosystems are being threatened by people and their activities

- That there are conflicts between different users of the coast, with each having their own views on coastal management

- The causes of coastal flooding and how to predict and prevent it

- The advantages and disadvantages of different coastal management strategies

This chapter looks at the coast, its landforms and the processes that produce them. Coastal ecosystems are very important. They are rich in biodiversity and resources. The exploitation of these resources is threatening their survival. This is just one of a number of conflicts occurring as development and conservation come face to face. There is a huge need for proper management of the coast.

▼ Figure 2.1: Factors affecting the coast

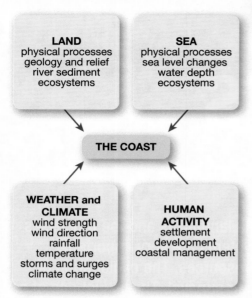

2.1 COASTAL PROCESSES

The coast is the transition zone between the land and the sea. The coastline is the actual frontier between the two. The coastal environment consists of two parts – **onshore** and **offshore** – located on either side of the coastline. The onshore zone can extend up to 60 km inland. The offshore zone reaches as far as 370 km out to sea.

The coastal environments of the world are very diverse. Figure 2.1 identifies the main factors causing this diversity. The diagram reminds us that coasts are the meeting point of not just land and sea, but also the atmosphere. Because of this, weather and climate are important factors. So, too, are the people who live and work along the coast.

The physical processes shaping the landforms of the coastline fall into two groups: marine (sea) and terrestrial (land).

CHECK YOUR UNDERSTANDING

Try to remember the four factor boxes in Figure 2.1 and their contents.

MARINE PROCESSES

Waves do much of the work of marine processes. They erode, transport and deposit materials. Waves are created by winds as they blow over the surface of the sea. It is friction between the wind and the water that sets waves in motion. The strength of waves depends on the strength of the wind. It also depends on the length of time and distance over which the wind has been blowing (the **fetch**).

As waves near the coast, they enter shallower water. Friction with the sea bed causes the wave to tip forward so that it eventually breaks. The resulting forward movement of water, called the swash, runs up the **beach** until it runs out of energy. The water then runs back down the beach under gravity. This is called the backwash.

▼ Figure 2.2: Constructive and destructive waves

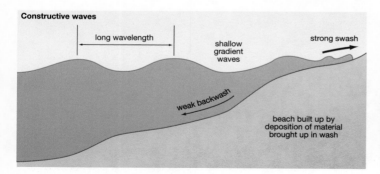

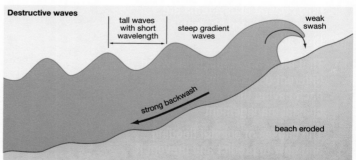

The balance between the swash and backwash of waves creates the difference between constructive and destructive waves (Figure 2.2). In constructive waves, the swash is stronger than the backwash. As a result, material is moved up the beach and much is left there (deposition). In destructive waves, the backwash is stronger. Material is dragged back down the beach (erosion) and moved along the coast by **longshore drift** (transport).

It is destructive waves that cause much of the erosion along a coast. They cut away at the coastline in a number of different ways:

■ hydraulic action – this results from the force of the waves hitting the cliffs and forcing pockets of air into cracks and crevices
■ abrasion – this is caused by waves picking up stones and hurling them at cliffs and so wearing the cliff away
■ corrosion (solution) – the dissolving of rocks by sea water.

Attrition is a process whereby the material carried by the waves becomes rounded and smaller over time as it collides with other material. It does not erode the coast as such but does form small pebbles and sand.

▶ Figure 2.3: The process of longshore drift

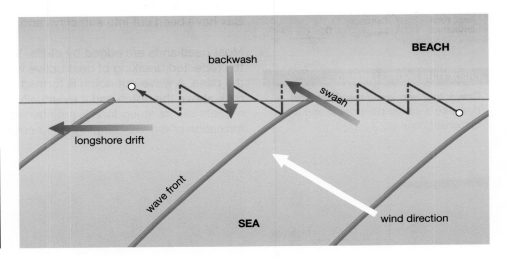

Once rocks and sand are detached from the cliff, waves can move them along the coastline for quite long distances. This process is known as longshore drift. Figure 2.3 illustrates how it works. The smaller the material, the lighter it usually is, and the further it is likely to be moved by waves. Eventually, the waves are unable to move so much material and the material will be deposited to create new landforms.

LAND PROCESSES

SKILLS ▶ REASONING

ACTIVITY

Explain how mass movement differs from erosion.

There are three main processes at work on the landward side of the coastline:

■ weathering – the breakdown of rocks which is caused by freeze-thaw and the growth of salt crystals, by acid rain and by the growth of plant roots
■ erosion – the wearing away of rocks by wind and rain
■ mass movement – the removal of cliff-face material under the influence of gravity in the form of rock falls, slumping and landslides.

2.2 COASTAL LANDFORMS

The interacting processes described in Part 2.1, together with four factors to be examined in Part 2.3, produce a variety of different coastal landforms. These landforms are broadly divided into those that result from coastal erosion and those from coastal deposition.

EROSIONAL LANDFORMS

▼ **Figure 2.4: Part of the Purbeck coast, southern England**

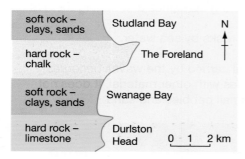

soft rock – clays, sands	Studland Bay	N
hard rock – chalk	The Foreland	
soft rock – clays, sands	Swanage Bay	
hard rock – limestone	Durlston Head	0 1 2 km

CHECK YOUR UNDERSTANDING

Check that you know the difference between a concordant coast and a discordant one.

By far the most common coastal landforms are the alternations of **headlands** and **bays**, which give many coastlines their irregular appearance. Destructive waves play an important role in their formation. The nature of the coastal rocks also plays a part. The direction in which rocks occur in relation to the coastline affects the resulting landforms. Coasts where the rock outcrops run parallel to the sea are called concordant coasts and often produce straighter coastlines. Coasts where the rocks outcrop at right angles to the sea are called discordant coasts and often produce headlands and bays. Weak rocks, such as clays, are easily eroded by the sea and also give rise to bays. Outcrops of more resistant rocks, that are able to withstand the destructive waves, protrude as headlands. Figure 2.4 shows a discordant coast in southern England. It shows outcrops of chalk and limestone forming the headlands, while Studland Bay and Swanage Bay have been cut into soft clays and sands.

Most headlands are edged by **cliffs**. Where cliffs rise steeply from the sea, the repeated breaking of destructive waves leads to them being undercut at the base. A wave-cut notch is formed (Figure 2.5). Undercutting weakens the rock above the notch and it eventually collapses. This means that the cliff face retreats but its steep face is maintained. The retreat of the cliff leads to the formation of a gently sloping wave-cut platform at the base.

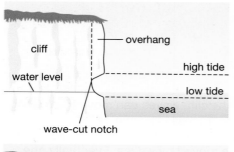

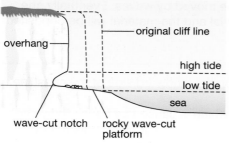

▲ **Figure 2.5: The formation of cliffs and wave-cut platforms**

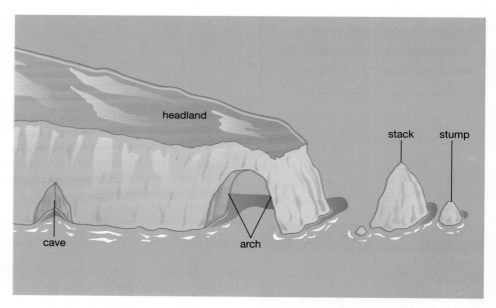

▲ **Figure 2.6: The formation of caves, arches, stacks and stumps**

Caves, **arches**, **stacks** and stumps also form on the sides of headlands as a result of constant attack on the headland's rocks by destructive waves (Figure 2.6). Any points of weakness in the headland's rocks, such as faults or joints, are attacked. The processes are largely hydraulic action and abrasion. This is likely to lead to the opening up of a cave. If the cave is enlarged and extends back through to the other side of the headland, possibly meeting another cave, an arch is formed. Continued erosion by the sea widens the arch. As the sea undercuts the pillars of the arch, the roof is weakened and eventually collapses. This leaves a stack separated from the headland. Further erosion at the base of the stack may eventually cause it to collapse as well. This leaves a small flat portion of the original stack, which is called a stump. It may only be visible at low tide.

▶ **Figure 2.7: Old Harry Rocks, southern England**

SKILLS INTERPRETATION

SKILLS INTERPRETATION

ACTIVITY

On Figure 2.7 identify:

■ a cave
■ an arch
■ a stack.

An excellent example of these erosional features is Old Harry Rocks (Figure 2.7) at the end of the chalk headland separating Swanage Bay from Studland Bay in southern England (Figure 2.4).

DEPOSITIONAL LANDFORMS

Depositional landforms are produced on coastlines where mud, sand and shingle accumulate faster than they can be moved away by the waves. This usually happens along stretches of coastline dominated by constructive waves (where the swash of water coming onto the beach is stronger than the backwash).

Beaches (Figure 2.8) are the most common depositional landform. They result from the accumulation of material deposited between the storm-beach area and low-tide marks. The sand, shingle and pebbles come from a number of sources. Much of it is material that has been eroded elsewhere and is being moved along the coast by longshore drift. Some comes from offshore as a result of waves picking it up from the sea bed and rolling it in towards the land. From the opposite direction, rivers feed mud and silt into the coastal zone via their estuaries (mouths). The deposition of this river material then takes place at the heads of sheltered bays.

▶ **Figure 2.8: Some beach features**

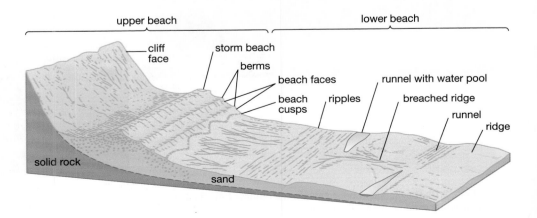

Many beaches show a number of very small features (Figure 2.8). For example, at the top end there can be a storm beach made up of large material thrown up during storm conditions. A series of small ridges, known as berms, mark the positions of mean (average) high-tide marks. Beach cusps – small semi-circular depressions – are formed by the movement of the swash and backwash of waves up and down the beach. In general, the finer the beach material, the more gently sloping the overall beach.

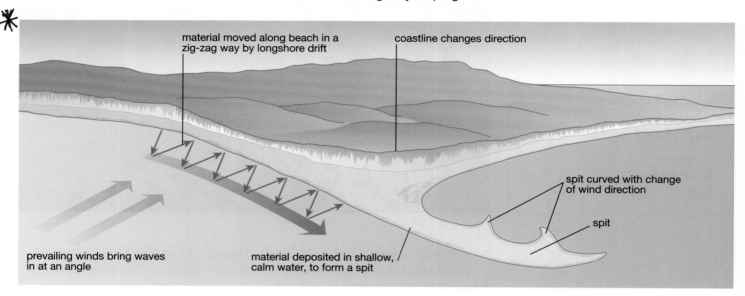

material moved along beach in a zig-zag way by longshore drift

coastline changes direction

spit curved with change of wind direction

spit

prevailing winds bring waves in at an angle

material deposited in shallow, calm water, to form a spit

▲ **Figure 2.9: The formation of a spit**

Spits are long narrow beaches of sand or shingle that are attached to the land at one end. They extend across a bay, an estuary or where the coastline changes direction. They are generally formed by longshore drift in one dominant direction (Figure 2.9). At the end of the beach, the material being transported by longshore drift is deposited. At a river estuary, the growth of the spit also causes the river to drop its sediment. This occurs mainly on the landward side (rather than the seaward side) of the spit and salt marshes may form. The waves and wind may curve the end of the spit towards the land.

If a spit develops in a bay, it may build across it and link the two headlands to form a **bar**. The formation of a bar is only possible if there is a gently sloping beach and no sizeable river is entering the bay. In this way, bars can straighten the coast and any water on the landward side is trapped to form a lagoon. An attractive example is found at Sotavento on the south coast of Fuerteventura, Canary Islands, Spain (Figure 2.10).

SKILLS ▶ CRITICAL THINKING

ACTIVITY

Do spits have their uses? Give examples.

▲ **Figure 2.10: The lagoon at Sotavento, Fuerteventura. Note the bar of fine, white sand enclosing the lagoon.**

▲ **Figure 2.11: The tombolo leading from Sharp Islands to Kiu Tao, Hong Kong**

SKILLS INTERPRETATION

ACTIVITY

Make an annotated sketch map of what is shown in Figure 2.11.

Tombolos are spits that have continued to grow seawards until they reach and join an island. A tombolo 1 km in length now joins the small island of Kiu Tau with the larger Sharp Island in Hong Kong (Figure 2.11). The formation of this tombolo is the result, not only of longshore drift, but also probably of offshore rocks and sand being rolled towards the shore.

Cuspate forelands are triangular-shaped accumulations of sand and shingle that extend seawards. It seems likely that many of them develop as a result of longshore drift occurring from two different directions. The head-on collision of the two drifts leads to the deposition of sediment and so to the formation of the foreland.

Sand **dunes** are depositional landforms in coastal areas. However, their formation is only indirectly related to coastal processes. Beaches are the source of sand which, when dry, is blown inland by the wind to form dunes. Over long periods of time, the wind blows the sand up into a series of ridges running parallel to the coastline. Gradually, the older ridges become colonised by vegetation and this helps to stabilise them (see Figure 2.23 on page 48).

SKILLS INTERPRETATION

ACTIVITY

Make a simple sketch, with notes explaining how the coastal landform shown in Figure 2.12 has been formed.

▶ Figure 2.12: A stretch of the Jurassic coast, southern England

SKILLS SELF DIRECTION

ACTIVITY

Find out why the Jurassic coast of southern England has been recognised as a World Heritage Site.

Many coastal landforms are put to good use by people. They offer opportunities that are exploited by various economic activities. For example, sandy beaches and stretches of fine scenery are exploited by tourism, as along the Jurassic coast of southern England (Figure 2.12). This is now designated as a World Heritage Site. Some of the world's most famous golf courses are laid out on sand dunes. Marinas are often built on the shores of estuary mouths sheltered by spits. Exposed coasts are used for wind farms. Nuclear and thermal power stations at the coast are further evidence of the energy industry exploiting these areas. More examples of human use of the coast follow in Part 2.6, and the resulting conflicts are explored in Part 2.7.

2.3 FACTORS AFFECTING COASTAL ENVIRONMENTS

Figure 2.1 on page 34 shows that there are other factors besides marine and land processes affecting the coast, particularly the coastline and its landforms.

GEOLOGY

As shown in Figure 2.4 on page 36, the difference between hard and soft rocks is a strong influence on the shape of the coastline. A coastline made up of weak rocks, such as clays and sands, will be easily eroded by destructive waves. Bays will be created. Coastlines of more resistant, harder rock will not be eroded so quickly.

Outcrops of more resistant rock along the coast often jut out into the sea as headlands. The difference between hard and soft rocks impacts on the shape and characteristics of cliffs (Table 2.1).

▶ Table 2.1: Contrasts between cliffs made of hard rocks and cliffs made of soft rocks

FEATURE	HARD ROCKS	SOFT ROCKS
Shape of cliffs	High and steep	Generally lower and less steep
Cliff face	Bare rock and rugged	Smoother; evidence of slumping
Foot of cliff	Boulders and rocks	Few rocks; some sand and mud

So, geology affects the coastline in two different ways:

- in plan view – headlands and bays
- vertically – the height and profile of cliffs.

CHECK YOUR UNDERSTANDING

What sort of rock would you expect to underlie a headland?

VEGETATION

In general, the longer a coastal landform (such as a sand dune) has existed, the greater the chances that it will be colonised by vegetation. In order to survive, vegetation has to cope with particular conditions, such as high levels of salt in both the air and soil. The major impact of vegetation is to protect and preserve coastal landforms. This is well shown in the cases of sand dunes (see Figure 2.23 on page 46) and mangrove swamps (see Figure 2.20 on page 44).

SEA-LEVEL CHANGES

One of the obvious effects of global warming and climate change is that low-lying coasts will be drowned by rising sea levels. This problem is made worse by the fact that many of the world's most densely populated areas are located on coastal lowlands.

In fact, rising sea levels are nothing new. During the Ice Age, sea levels also changed but to a much greater extent. They fell as more and more of the world's water was locked up in ice sheets and glaciers. The sea levels rose again as the ice sheets and glaciers melted.

▶ **Figure 2.13: A fjord in Norway**

A rising sea level leads to a submerged coastline. The main features are rias (drowned river valleys) and fjords (drowned glacial valleys, as shown in Figure 2.13).

▶ **Figure 2.14: A raised beach in Scotland**

ACTIVITY

Of these three factors – geology, vegetation and sea-level changes – which one do you think is the most significant? Give your reasons.

CHECK YOUR UNDERSTANDING

Can you pick out the raised beach and relict cliffs in Figure 2.14?

An emerging coastline is associated with a falling sea level. The most common landforms are **raised beaches** (Figure 2.14). These are areas of wave-cut platform and their beaches are now found at a level higher than the present sea level. In some places, relict cliffs with caves, arches and stacks are found where there are raised beaches.

HUMAN ACTIVITIES

Human activities can have significant effects on the coastal environment. At this stage, we need only identify the main ones. They will be explored and explained in more detail later in this chapter. The main human activities are outlined below.

- Settlement – coastal lowlands have attracted people and their settlements worldwide throughout history. Many of the world's most densely populated areas are located on the coast.

SKILLS INTERPRETATION

ACTIVITY

Look closely at Figure 2.15. Identify the different land uses that are likely to be involved in this coastal development.

▶ Figure 2.15: Recent coastal development, Qatar

SKILLS CRITICAL THINKING

ACTIVITY

Discuss, in groups, why the coast is attractive for human settlement and development.

- Economic development – people have taken advantage of the economic opportunities that the coast offers, such as land for agriculture and industry. Fishing and the chance to trade either along the coast or overseas have led to the building of ports and harbours (Figure 2.15). As illustrated at the end of Part 2.2, the coast is often used for **tourism** and by the energy business. All these activities and others impact on the coastal environment.
- Coastal management – for many centuries, people have tried to control the coastline. For example, building sea walls and groynes helps to protect stretches of coastline from high rates of erosion or deposition (Figure 2.33 on page 57).

As a result of these and other human activities, the natural landscapes and features of the coast can be greatly changed. Sometimes the actual shape of the coastline is altered.

2.4 COASTAL ECOSYSTEMS OF THE WORLD

Given the large amounts of development and settlement in coastal areas of the world, it is easy to forget that the coast is home to a variety of ecosystems. In their natural state, coasts can be very rich in **biodiversity**. In this section, we will look at four different ecosystems. Two of them, coral reefs and mangroves, are found in the tropics. The other two, salt marshes and sand dunes, are common across the world.

CORAL REEFS

Coral reefs are a unique marine ecosystem. They are built up entirely of living organisms (Figure 2.16). Reefs are huge deposits of calcium carbonate made mainly of corals. Their global distribution is shown in Figure 2.17. It is mainly controlled by four factors:

- temperature – coral growth needs a minimum water temperature of 18°C; they grow best between 23°C and 25°C
- light – needed for the coral to grow; because of this, corals grow only in shallow water

▲ Figure 2.16: Section of a coral reef, St Lucia

- water depth – because of the need for light, most reefs grow where the sea is less than 25 metres deep
- salinity – since corals are marine creatures, they can only survive in saltwater.

At a local level, other factors affect where coral reefs develop:

- wave action – corals need well oxygenated saltwater; this occurs in areas of strong wave action
- exposure to air – corals need oxygenated water, but if they are exposed to the air for too long they die
- sediment – corals need clear, clean water; any sediment in the water blocks their normal ways of feeding and reduces the amount of light.

Coral reefs with the highest biodiversity occur in South-East Asia and northern Australia (Figure 2.17). The Great Barrier Reef in Australia is renowned for its great biodiversity, and also its extent.

ACTIVITY

Look closely at Figure 2.17 and make notes of the main features of the global distribution of coral reefs.

▶ Figure 2.17: The global distribution of coral reefs

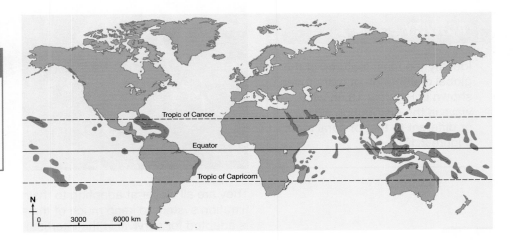

MANGROVES

Mangroves are most common in South-East Asia (Figure 2.18) where they were thought to originate before spreading around the globe. Today, most mangroves are found within 30° latitude of the equator, but a few hardy types have adapted to temperate climates. They reach as far as the North Island of New Zealand.

Mangroves literally live on the coastline – they have one 'foot' on the land and the other in the sea. Because they grow in the intertidal zone, they live in a constantly changing environment. They are regularly flooded by the sea. At low tide, especially during periods of high rainfall, there may be floods of fresh water. This quickly alters the salt levels, as well as temperatures. Mangroves are not only able to survive these changing water conditions, they can cope with great heat and choking mud.

Despite these environmental difficulties, the mangrove ecosystem is among the most successful ecosystems on Earth. South-East Asia also has mangroves with the highest biodiversity in the world. There are many species of mangrove. They range in size from small shrubs to trees over 60 m in height.

▶ Figure 2.18: The global distribution of mangroves

▶ Figure 2.19: Mangroves on the coast of Thailand

SKILLS ▶ INTERPRETATION

ACTIVITY

Identify the main features of the global distribution of mangroves shown in Figure 2.18. In what ways is the distribution similar or different to that of coral reefs (Figure 2.17)?

SKILLS ▶ CRITICAL THINKING

ACTIVITY

Discuss, in groups, what you think is special about mangroves.

They are all clever at adapting to their environment. Each mangrove has a filtration system to keep much of the salt out and a complex root system that is adapted for survival in the intertidal zone. Some have snorkel-like roots that stick out of the mud to help them take in air, as in Figure 2.19; others use prop roots or buttresses to keep their trunks upright in the soft sediments at the tidal edge.

It is these roots that trap mud and sand, and eventually build up the intertidal zone into land. At the same time, the mangrove is colonising new intertidal areas. The fruits and seedlings of all mangrove plants can float. As they drift in the tide away from the parent trees, they become lodged in mud where they begin to grow. So, a new area of mangrove takes root.

SALT MARSHES

Salt marshes occupy a midway location between mudflats that are permanently submerged by water and terrestrial (land) vegetation lying above the high-tide mark. Like mangroves, they are an ecosystem of the intertidal zone.

Coastal salt marshes develop in locations sheltered from the open sea, namely at the heads of bays and in estuaries. Since estuaries are where a river meets the sea, the water is brackish (partly salty and partly fresh). In bays, the water is salty.

▶ Figure 2.20: Temperate salt marsh zones

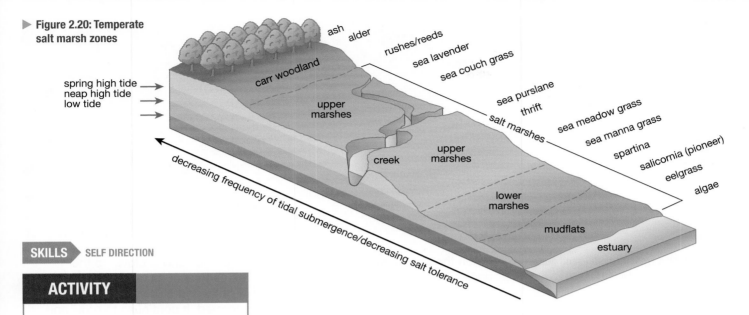

SKILLS ▶ SELF DIRECTION

ACTIVITY

Find out the difference between spring tides and neap tides. What is the significance of the difference in terms of salt marsh ecosystems?

Salinity (how salty the water is), together with the frequency and extent of flooding of the marsh, determine the types of plants and animals found there (Figure 2.20). In some cases, the low marsh zone floods twice daily while the high marsh floods only during storms and unusually high tides. These varying environmental conditions result in different types of plants and animals being found in different parts of the same marsh area.

▶ Figure 2.21: A typical area of salt marsh, UK

SKILLS ▶ INTERPRETATION

ACTIVITY

Which of the salt marsh zones is shown in Figure 2.21? Refer to Figure 2.20 to see the possible choices. Give reasons for your choice of zone.

Salt marshes are criss-crossed by meandering creeks, which allow tidal water to drain in and out (Figure 2.21). The creeks slow tidal energy and the marsh plants slow down wave energy. As a result, there is an almost continuous deposition of silt and mud. Over time, this means that the salt marsh gradually extends seawards.

COASTAL SAND DUNES

Coastal sand dunes are accumulations of sand shaped into mounds and ridges by the wind. They develop best where:

■ there is a wide beach and large quantities of sand
■ the prevailing wind is onshore (from the sea to the shore)
■ there are suitable locations for the sand to accumulate.

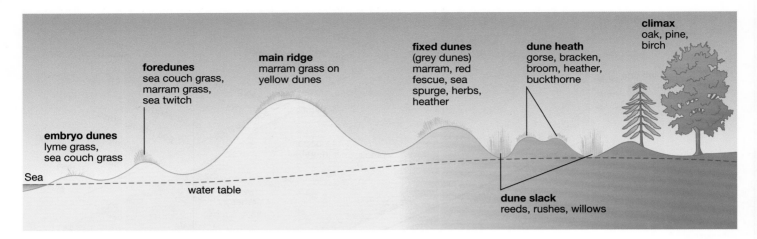

▲ Figure 2.22: A typical coastal sand dune cross-section in northern Europe

When the beach dries out at low tide, some of the sand is blown to the back of the beach by the onshore wind. The sand accumulates there, often around a small obstacle such as a piece of driftwood or dry seaweed. As the accumulation grows, a small sand dune is formed. It continues to grow and becomes more stable. Another dune may eventually develop on the seaward side of the original dune. The original dune is now further inland and relatively sheltered from the prevailing onshore wind. If this sequence continues, a series of dunes will develop in the form of ridges running parallel to the shore (Figure 2.22).

Over time, the ridges of the dunes will be colonised and 'fixed' by vegetation. The older the ridge and the further inland, the greater the vegetation cover. The first plants, such as sea twitch and sea couch grass, see Figure 2.23, have to cope with the following conditions:

■ salinity
■ a lack of moisture as sand drains quickly
■ wind
■ temporary submergence by wind-blown sand.

▶ Figure 2.23: A belt of embryo and foredunes, Sardinia

SKILLS   REASONING

ACTIVITY

How and why does the vegetation change as you move inland across a belt of sand dunes?

Once some plants become well established, environmental conditions will improve and other plants begin to appear. Eventually, in temperate areas, dune heath will be established (Figure 2.22). This whole sequence is known as a **plant succession**.

2.5 COASTAL ECOSYSTEM CHARACTERISTICS

In Part 2.4 we looked at four main coastal ecosystems. Despite their distinctive locations wholly or partially in saltwater, they function in the same way as land-based ecosystems, such as the tropical rainforest or the tundra. They have their biotic components – the living parts (plants and animals). They also have their abiotic components – parts of the ecosystem that are not living, but are essential to life (climate, minerals).

All ecosystems survive by nutrient cycling, that is by the transfer of nutrients within them. This involves circulating minerals around three stores. The first is the nutrient source (in terrestrial ecosystems this is the soil). Nutrients are then transferred to support the **biomass** (the living matter of the ecosystem). As some of the biomass dies, nutrients are released as **litter**. This litter subsequently supplies new nutrients to the nutrient source and so completes the cycle and a new one starts.

It is relatively easy to understand how the nutrient cycle works in three of the coastal ecosystems – mangrove, salt marsh and sand dunes. In all of them, the 'soil' store lies mainly in either mud or sand. Their biomass stores, particularly the plants, are easy to see – even though, in the case of mangrove and salt marsh, some may be below the water line. The same applies to their litter. Much of it collects and degrades in water. Clearly, the sand dune ecosystem is slightly different here because the nutrient transfers take place on dry land.

CHECK YOUR UNDERSTANDING

Check that you know the difference between biomass and litter.

▶ Figure 2.24: The nutrient cycle of coastal ecosystems

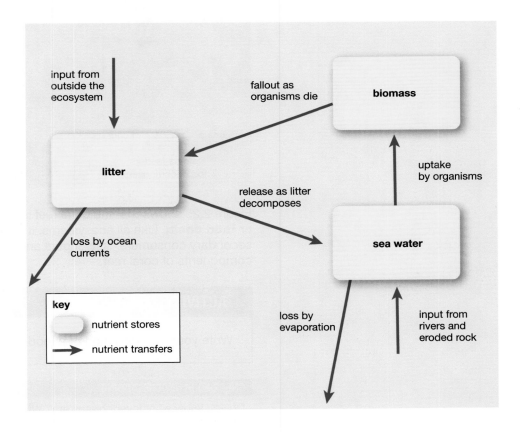

It is more difficult to understand how the nutrient cycle model applies to the coral reef ecosystem, particularly to the coral that is such an important component. The three stores are listed below.

■ Sea water – this is the 'soil' of a coral reef ecosystem. The supply of nutrients in the sea water is maintained by on-shore ocean currents. Abiotic nutrients are also supplied by rivers entering the sea. They are filtered out and absorbed by the coral and other small organisms and so the coral reef grows. Tiny algae living within the coral convert the energy in sunlight into energy for the coral.
■ Biomass – the biomass of the coral reef has three main components: the coral itself (the coral polyps are living creatures); the great variety of seaweeds attached to the coral; and the huge diversity of fish, crustaceans and invertebrates taking advantage of opportunities provided by the reef.
■ Litter – this is made up of dead coral but more importantly by the remains of dead fish, invertebrates and seaweeds. Litter is also moved around by tidal and ocean currents. It enriches the sea water with nutrients which are then filtered out by the coral.

▶ Figure 2.25: The possible food web of a coral reef ecosystem

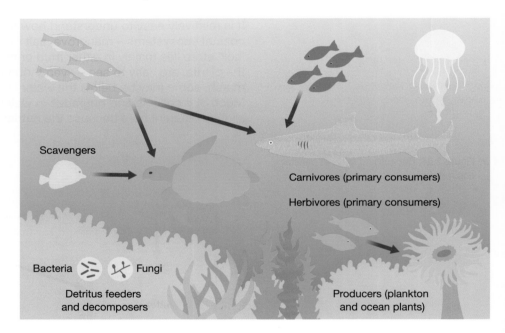

Figure 2.25 shows that the coral reef is home to a well-organised food web or **food chain**. Like all ecosystems, it comprises producers, primary and secondary consumers, scavengers and decomposers. These are the biotic components of coral reef.

ACTIVITY

Write your own definition of a food web.

CHECK YOUR UNDERSTANDING

Do you know what invertebrates are? What is the difference between a food web's producers and its consumers?

2.6 COASTAL ECOSYSTEMS UNDER THREAT

In this section, we continue to look at the same four ecosystems but in terms of their value to people and how this, in turn, threatens their existence. All ecosystems offer people a range of opportunities. We refer to these opportunities as goods and services. Table 2.2 shows the main goods (material resources that can be extracted and used) and services (general benefits and advantages) provided by coastal ecosystems.

▶ Table 2.2: Goods and services provided by coastal ecosystems

GOODS	SERVICES
Fish and shellfish	Protection from storms
Fishmeal and animal feed	Harbours
Seaweed for food and industrial use	Shelter
Salt	Recreational opportunities
Land for settlement and farming	Biodiversity and wildlife habitats
Construction materials such as sand and timber	Natural treatment of wastes

CHECK YOUR UNDERSTANDING

Be sure you understand the difference between the goods and services provided by an ecosystem.

CORAL REEFS

The value of coral reefs lies in:

- their biodiversity – within the Great Barrier Reef there are 700 species of coral, 1500 species of fish and 4000 species of mollusc
- the protection they give to low-lying coasts from the impact of tropical storms
- their rich fish stocks – they supply the basic food requirements of many developing countries
- their appeal to tourists and the recreational opportunities they offer such as snorkelling and scuba diving; over 150 million people each year take holidays in areas with coral reefs.

Coral reefs are easily stressed by human actions. Any contact with the human body is likely to kill a coral immediately around the point of contact. Reefs are also threatened by pollution, overfishing and the quarrying of coral for building stone. If the stress persists, the death of the reef soon follows. Figure 2.26 shows the sequence of coral reef decline that follows from development of the coastal area nearby. Development involves the spread of urban areas, the growth of a fishing industry, the coming of commercial farming and the rise of tourism. These have a range of secondary impacts that eventually lead to the decline and possible death of the coral reef.

A recent survey of the world's coral reefs showed that 27 per cent were highly threatened by human activities. Another 31 per cent were classified as being under 'medium threat'. But coral reefs are also being threatened by coral bleaching. This is the result of rising water temperatures associated with global warming.

▶ Figures 2.26: A model of coral reef decline

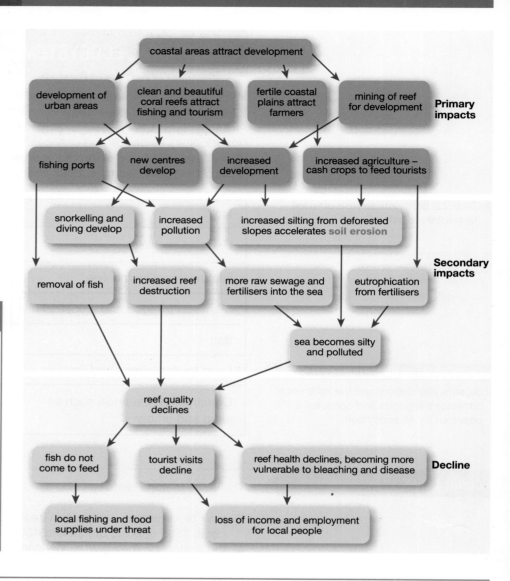

ACTIVITY

SKILLS ▷ INTERPRETATION

Look at Figure 2.26 and suggest reasons for the following links:

■ increased development, which leads to increased pollution
■ sea becomes polluted, which leads to decline in reef quality
■ reef health declines, which leads to loss of income and loss of employment.

MANGROVES

▼ Figure 2.27: The global destruction of mangroves

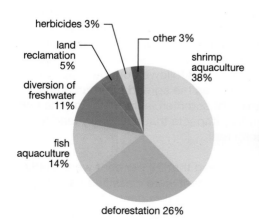

Mangroves are valuable nurseries for fish and crustaceans (shellfish), and are rich in wildlife. Mangrove roots, which are exposed at low tide, trap silt and help to create new land. Mangrove timber provides fuel and building material. However, perhaps the greatest value of mangroves in this age of rising sea levels is the protection from storm surges that they give to low-lying coastal areas. The World Conservation Union compared the death tolls in two Sri Lankan villages of the same size that were hit by the 2004 tsunami. Two people died in the settlement protected by dense mangrove, while up to 6000 people died in the village without such protection.

SKILLS ▷ ANALYSIS (NUMERICAL)

ACTIVITY

In class, discuss the arguments for and against protecting mangroves from human use.

It is widely believed that mangrove swamps are disease-ridden. For this reason, they are being cleared at a fast rate. Figure 2.27 shows the causes of this destruction at a global scale. Just over half of the mangrove swamps are cleared to make way for aquaculture – the farming of fish and shrimps. A quarter are cleared (deforestation) to provide timber for fuel and building purposes. Land reclamation is undertaken to provide sites for building tourist hotels and other amenities. The diversion of fresh water is intended to meet the needs of tourists and expanding agriculture. Farming also requires the application of herbicides to prepare cleared areas for cultivation.

CASE STUDY: MANAGEMENT OF THE MANGROVE

Bangladesh is one country that understands the value of the mangrove. It is a low-lying country. Its long coastline is vulnerable to tropical storms and storm surges. Bangladesh also has a huge population and a shortage of land. The mean population density is around 1000 persons per km^2. Up to 25 million people live less than 1 metre above sea level.

As part of its Coastal Zone Policy, the Bangladesh government is taking advantage of the fact that mangroves trap sediment and stabilise shores. By deliberately planting mangroves on delta sediments washed down from the Himalayas, it has gained over 120 000 hectares of new land in the Bay of Bengal. The plantings are relatively new, but there have been mangroves here for as long as the Ganges, Brahmaputra and Meghna Rivers have been draining into the bay. The vast tidal mangrove woodland is known as the Sundarbans – it literally means 'beautiful forest'. Today, it is the largest surviving single tract of mangroves in the world (Figure 2.28).

The management of the Sundarbans creates new land and protects Bangladesh from coastal flooding, while also permitting local people to make use of its resources. The only real threat is the clearance

of some mangroves to allow for the aquaculture of shrimps and fish. Shrimps are a valuable export and fish are an important food to help feed a large and hungry population.

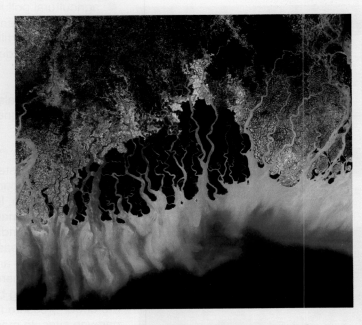

▲ Figure 2.28: Satellite image of the Sundarbans, Bangladesh

Unfortunately, the popular perception of mangroves today continues to undervalue them. For this reason, they are under threat, particularly from various forms of development.

ACTIVITY

Make a list of the benefits provided by mangroves.

SALT MARSHES

Salt marshes may appear to be of little obvious value. But in fact, they have a number of valuable uses, such as:

- collectors of silt and organic matter
- nursery areas for fish and crustaceans
- protection against wave erosion and sea-level rises.

▼ Figure 2.29: Industrial development on a salt marsh, UK

Despite this natural value, many salt marshes are among the most abused and therefore the most threatened ecosystems in today's world. Specific threats include:

- reclamation to create farmland and sites for industrial and port developments – this is based on the perception that marshes are wasted spaces that need to be put to some good use
- industrial pollution – particularly of water, as many marshes occur in estuaries which are favoured as sites for ports, power stations and oil refineries (Figure 2.29)
- agricultural pollution – heavy applications of fertilisers and pesticides on adjacent farmland lead to eutrophication (an increase in the concentration of chemical fertilisers in an ecosystem) of marshland waters
- pressure from developments such as marinas and other recreational facilities.

SKILLS REASONING

ACTIVITY

Suggest reasons why salt marsh areas are attractive to industry.

Salt marshes are also threatened by changes associated with global warming such as more frequent storms and higher water levels.

COASTAL SAND DUNES

▼ Figure 2.30: An area of degraded sand dune, UK

Of the four ecosystems, coastal sand dunes are probably the least threatened at a global level, perhaps because they have little to offer people other than coastal protection. In the UK, however, they are put at risk by people using them as recreational spaces. Various forms of recreation – such as trail biking and horse riding – are doing great damage. Sand dunes are delicate ecosystems and easily disturbed (Figure 2.30). Disturbance often leads to a loss of vegetation and to blow-outs (depressions in sand dunes caused by wind erosion). Where sand dunes are close to urban and industrial areas, they are at risk of being built over. Because of their nearness to the coastline, they are under pressure from tourism to provide amenities such as golf courses and caravan sites. In many parts of the world, areas have been planted with trees to help stabilise mobile dunes. The net effect has been to destroy the coastal sand dune ecosystem.

2.7 COASTAL CONFLICTS

Coastal environments are of great importance to people. For this reason, they need to be carefully managed so that the same opportunities are still available for future generations.

THE COASTAL SYSTEM

The coast, like drainage basins (see page 5) and ecosystems (see page 47), may be thought of as a system (Figure 2.31). To understand what this means, think of the sea as an invisible box. Things enter that box (inputs) from both sides of the coastline, and either remain there (stores) or pass

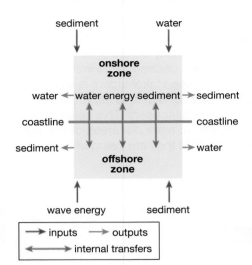

sediment   water

onshore zone

water ← water energy sediment → sediment

coastline ——————— coastline

sediment ← ↕ ↕ ↕ → water

offshore zone

wave energy   sediment

→ inputs   → outputs
←——→ internal transfers

▲ **Figure 2.31: The coastal system**

through it (flows or transfers) and eventually leave it (outputs). One input is sediment. This comes from rivers and the weathering and erosion of cliffs. The stores of sediment are beaches, spits, forelands and sand dunes. The transfer of sediment is mainly the movement of sand and shingle along the coast by longshore drift. The loss of sediment from the coast to the open sea is an output.

Where there is a balance between inputs and outputs, the system is said to be in equilibrium. This balance can, however, be upset by natural changes. As we saw in Part 2.6, human activities can seriously damage coastal ecosystems. In this section, we will see that most human activities lead to a range of environmental impacts and conflicts.

ACTIVITY

Compare the coast and the drainage basin as systems.

CONFLICT BETWEEN DEVELOPMENT AND CONSERVATION

Each of the threats considered in Part 2.6 creates a specific conflict. Many conflicts are about human needs versus the well-being of coastal ecosystems. They all raise the basic question – should those ecosystems be protected and conserved? Or should people make the fullest use of their resources and opportunities, and exploit them? The overriding coastal conflict is between conservation and development.

ACTIVITY

Debate, in class, whether or not coastal development should be given a higher priority than coastal conservation.

CONFLICTS BETWEEN COASTAL USERS

The competition between development and conservation is just one source of conflict in coastal areas. There are many other conflicts between the users (sometimes called stakeholders) of the coast. These users are competing because of their particular needs. They all have their own views about the way the coast should be used and managed.

Who are the main users of coastal areas and what are their special needs? The users include:

SKILLS CRITICAL THINKING

ACTIVITY

Which **two** of the coastal stakeholders listed on the right do you think are most in conflict? Give your reasons.

- local residents – good choice of housing; clean environment
- employers – access to labour; space for shops, offices and factories
- farmers – well-drained land; shelter from strong onshore winds
- fishermen – harbours; unpolluted waters
- port authorities – harbours and space for port-side services and terminals
- transport companies – good roads and terminals such as ports and airports
- tourists – beaches; hotels; recreational amenities; heritage sites
- developers – greenfield sites.

The conflicts that occur in coastal areas can also be a problem on the seaward side of the coastline and between different users of the inshore waters. This is well illustrated in the waters of Lyme Bay on the south coast of England. Here there are at least six stakeholders or users of its waters (Figure 2.33). It is important to note that one of these stakeholders is wildlife. Here, as on land, the only sensible solution to the conflict is to separate the different uses by allocating specific areas to particular users. This solution is not as simple as it might seem. It is quite difficult to mark out areas of the sea. It is also difficult to police these areas to check that they are being used as they should be.

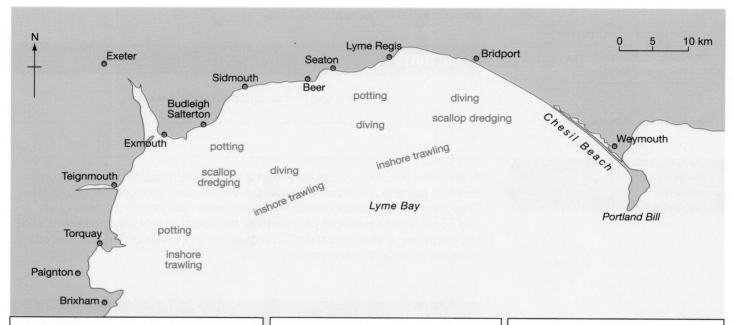

Wildlife lovers
The key habitats are the rocky reefs and the areas that support seagrass and maerl. These act as a nursery ground for commercially valuable invertebrates and fish, which provide food for large numbers of seabirds, such as auks, divers, gulls, terns and grebes. All three habitats are sensitive to physical damage.

Scallop dredgers
Boats operating out of the bay's ports bottom-trawl for scallops in inshore waters. The industry generates around £180 000 a year. Dragging heavy fishing gear across the seabed causes high levels of physical damage, stirs up sediment and smothers sensitive species. This is likely to have long-term negative effects on all the other users of the bay. If it were to be continued to the sand and gravel areas close to the shore, the impact would be even worse.

Trawlers
Three areas in Lyme Bay are important to trawlers: the central area, where red mullet, skate, squid and cuttlefish are found between June and November; a winter run of lemon sole from Portland Bill moving west across the bay; and sprat and herring in the late summer in the southwestern part. Fishing gear has an adverse impact on habitats.

Potters
Crab and lobster potting in the main reef areas generates nearly £180 000 a year for the local economy, but is incompatible with scallop dredging because pots can get tangled in the lines of a scallop dredger. Female edible crabs travel far along the coast to spawning grounds and it is important to protect their migration routes from damage.

Divers and anglers
Diving is a low-impact activity and generates about £87 000 a year, mainly through dive companies in Torbay, Exmouth and Lyme Regis visiting the wreck sites and rocky reef area. Angling trips are made across the bay for species such as cod and bass. Angling generates nearly £250 000 a year for the local economy. The impact on marine habitats is low, but fishing lines may snag sensitivespecies such as pink sea fans.

Tourists
Tourism contributes to the economy of the entire southwest region, as well as that of local communities. Sightseeing, sailing, water-skiing, jet-skiing, power-boating, windsurfing and wildlife watching are all popular but depend on a healthy, attractive environment.

▲ **Figure 2.32: Stakeholders in Lyme Bay, southern England**

2.8 COASTAL FLOODING

There is a difference between the gradual retreat of a coastline by erosion and the flooding of a low-lying coastline by occasional, abnormally high tide levels. Storm surges are the greatest flood threat. These are caused by very low air pressure, which raises the height of the high-tide sea. Strong onshore winds then drive the 'raised' sea towards the coast and are capable of breaching coastal defences and flooding large areas. Tsunamis generated by earthquakes can also lead to widespread coastal flooding as in the 2004 Indian Ocean tsunami (see the Sri Lanka case study).

Storm surges and tsunamis are periodic events. However, there are some stretches of coastline where the risk of flooding is both constant and increasing. This is the case with the city of Venice at the head of the Adriatic Sea in Italy. Global warming, and its associated rise in sea levels, is also increasing the coastal flood risk in many parts of the world.

DID YOU KNOW?

For more on tsunamis, see Sri Lanka case study (page 59) and Part 3.3 (page 69).

REDUCING THE COASTAL FLOODING RISK

The risk of coastal flooding may be reduced in two ways: by prediction and prevention. Prediction involves the following.

- Looking back at historic records and identifying those areas that have been flooded most often and most seriously, i.e. defining the high-risk areas.
- Relying on an accurate forecasting of possible hazard events. In the case of coastal flooding, this means forecasting the approach of tropical storms and their associated storm surges (see Part 3.8, page 87) and the occurrence of earthquakes and possible tsunamis (see Part 3.3, page 69). In both instances, the forecasting should not only say when and where the event will occur, but also predict its scale and strength and the likelihood of damage and death.

Prevention involves actions that reduce or even remove the risk of coastal flooding. Possible actions include:

- building flood defences along those stretches of the coast most at risk
- building emergency centres where people can be safe from the flooding
- removing housing and human activities from high-risk areas
- planning new development that avoids high-risk areas and designing new buildings able to cope with low levels of flooding
- installing advanced warning systems of possible flooding; clearly this depends on good forecasting
- educating local people on what to do once warning systems have been activated.

CHECK YOUR UNDERSTANDING

Be sure you know the difference between prediction and prevention. Give examples of **both**.

2.9 COASTAL MANAGEMENT STRATEGIES

Coastal management in many parts of the world today is based on identifying coastal cells. These are sections of the coast which are self-contained in terms of the movement of sediment. Coastal cells are mini-systems (see Figure 2.31 on page 53).

Coastal management is about two things. The first is resolving the conflicts, as we saw in Part 2.7, between different users of the coast and between those users and the well-being of coastal ecosystems. The second is taking action to manage big changes that threaten long stretches of the coast. These changes can present risks. Two major risks are the risk of coastal erosion, and the risk of coastal flooding. The two risks are related.

COASTAL EROSION AND RETREAT

Coastal erosion is quite natural, and in most places, it is unspectacular. However, some stretches of coast are eroding at alarming speeds. For example, at Holderness on the north-east coast of England, the cliffs (20–30 m in height) which are made up of soft sands, gravels and clays are retreating at a rate of 1 m per year – occasionally up to 10 m per year. Over the last 2000 years, the coastline has been pushed back 4 km.

COASTAL FLOODING

Coastal flooding can result from coastal erosion. It can also occur along low-lying coasts as a result of occasional, abnormally high sea levels. Perhaps more threatening are the periodic events, such as storm surges and tsunamis (see Chapter 3). But the greatest flooding risk is global warming and the rise in the global sea level.

SKILLS REASONING

ACTIVITY

Explain how coastal erosion and coastal flooding are related.

Over the last 150 years, people have tried to protect valued stretches of the coast from erosion and flooding. Much concrete, rock and timber has been used in trying to win the endless battle against the sea. All the management strategies used so far have involved either hard engineering or soft engineering.

HARD-ENGINEERING MANAGEMENT

Hard engineering involves building some type of sea defence, usually from rocks or concrete. It aims to protect the coast from erosion and the risk of flooding by working against the power of waves. Figure 2.34 illustrates some techniques used at the foot of cliffs and on beaches. Each has its strengths and weaknesses. For example, rip-rap is effective and cheaper to install than either sea walls or revetments. However, it may shift in heavy storms and be undercut by the backwash of waves. Hard engineering as a whole has several disadvantages.

SKILLS INTERPRETATION

ACTIVITY

Which type of hard engineering shown in Figure 2.34 would be best for:

- reducing the effects of longshore drift
- protecting the base of a cliff from erosion?

- Most structures are expensive to build and maintain – to repair a sea wall can cost up to £3000 a metre.
- Effective defence in one place can have serious consequences for a nearby stretch of coastline, particularly in the direction of longshore drift. For example, groynes trap beach material that is being moved by longshore drift on their upstream side. Downstream from the groyne, the lack of beach material increases the exposure of the coast to the forces of erosion.
- Defence structures such as sea walls, gabions and rip-rap cannot keep pace with rising sea levels.
- Structures can spoil the natural beauty of a coastline.

▶ **Figure 2.33: Examples of coastal hard engineering**

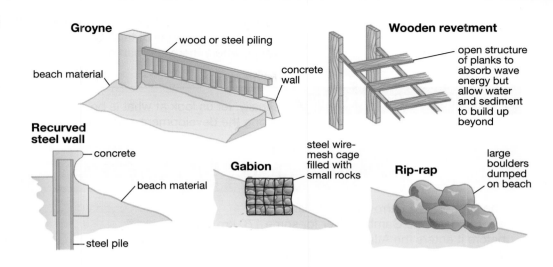

SOFT-ENGINEERING MANAGEMENT

▲ **Figure 2.34: Beach replenishment in progress, UK**

Soft engineering tries to work with natural processes. It makes use of elements of the coastal system, such as beaches, sand dunes and salt marshes. The following are examples of soft engineering.

■ Beach replenishment – pumping or dumping sand and shingle back onto a beach to replace eroded material (Figure 2.34).
■ Building bars – underwater bars located just offshore to reduce wave energy.
■ Fencing, hedging and replanting vegetation – this helps to preserve a beach or sand dune by reducing the amount of sand that is blown inland. Even more effective is the wholesale rehabilitation of degraded coastal ecosystems.
■ Cliff regrading – the angle of a cliff is reduced so that it is not so steep. This reduces the likelihood of cliff retreat by mass movement (a large part of the cliff suddenly falling down as a result of water erosion).

All of these actions are used to absorb wave and tidal energy. They do not disfigure the natural appearance of the coast. They are more environmentally-friendly than hard-engineering solutions. Soft-engineering strategies are generally much less expensive than hard-engineering ones.

ACTIVITY

Summarise the difference between hard- and soft- engineering approaches to coastal management.

MANAGED RETREAT

In recent years, another type of coastal management has appeared. It is probably a response to the rising sea levels of global warming. Some see managed retreat as yet another form of soft engineering. Others see it as a third form of coastal management. It involves abandoning existing coastal defences and allowing the sea to flood inland until it reaches higher land or a new line of coast defence. Allowing low-lying coastal areas to flood and develop into salt marshes produces a good natural defence against storms. It also increases the amount of salt marsh – an increasingly scarce and threatened ecosystem. It is a relatively cheap method of coastal defence.

The main cost is compensating people for the loss of 'drowned' homes and livelihoods. Because of this cost, managed retreat is not suitable in coastal areas where there is urban development and high quality farmland.

COASTAL MANAGEMENT IN ACTION

Finally, let us look at what is being done in three countries at different points along the development pathway.

CASE STUDY: COASTAL MANAGEMENT IN THE GAMBIA

The Gambia is a narrow finger of land lying either side of the Gambia River for some 170 km before it enters the Atlantic Ocean (Figure 2.35). With an area of just 11 000 km², it is one of the smallest states in the world. Surrounded on three sides by Senegal, this former British colony has an open ocean coastline of only 80 km. However, the coastline is dominated by the wide estuary of the Gambia River. This adds another 200 km of sheltered coast.

The coastline to the south of the estuary has been suffering from coastal erosion for a long time, but the rate of erosion has increased significantly over the last three decades. This is attributed to a slight rise in sea level, which, in turn, is linked to global warming. The sea-level change is a double blow for The Gambia. It stands to lose:

■ most of its capital city, Banjul, located at the mouth of the Gambia River (Figure 2.35)
■ much tourist infrastructure located close to the coast south of the river. This is the main tourist area. So, for a country so heavily dependent on tourism, coastal erosion is fast becoming a disaster.

For a developing country, such as The Gambia, coastal management is something of a luxury. It can do little to protect itself from the risks of coastal erosion and flooding.

A project, partly funded by the United Nations, is now investigating ways of enhancing the resilience of threatened coastal areas and communities. At least it is making the government and people more aware of the seriousness of the situation. A range of possible actions to be taken immediately has been outlined. The problem is that while some of those actions are relatively simple technologically, they are nonetheless costly. The only hope for The Gambia is to find external sources of funding. This is not easy for a country that has a reputation for corrupt government.

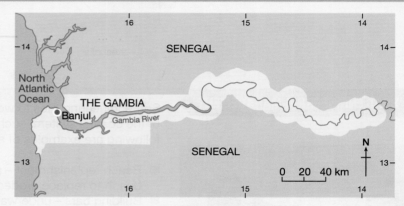

▲ Figure 2.35: The Gambia and the Gambia River

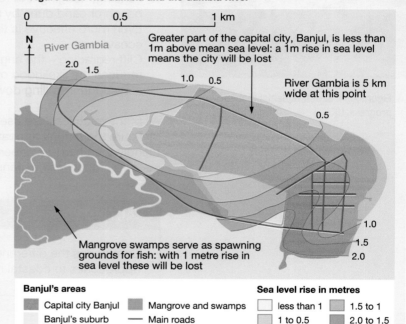

Greater part of the capital city, Banjul, is less than 1m above mean sea level: a 1m rise in sea level means the city will be lost

River Gambia is 5 km wide at this point

Mangrove swamps serve as spawning grounds for fish: with 1 metre rise in sea level these will be lost

Banjul's areas
Capital city Banjul
Banjul's suburb
Mangrove and swamps
— Main roads

Sea level rise in metres
less than 1
1 to 0.5
1.5 to 1
2.0 to 1.5

▲ Figure 2.36: The impact of sea-level rise in Banjul

ACTIVITY

How might you promote the case of The Gambia for 'outside help' with coastal management?

CASE STUDY: COASTAL MANAGEMENT IN SRI LANKA

Sri Lanka is an island state with a 1560 km-long coastline. Nearly half the country's population lives in the coastal zone (Figure 2.37). The most densely populated and urbanised part is the south-west coast between Colombo and Galle. The coast offers tourism which plays an important part in the Sri Lankan economy.

Problems include:

- pollution of inshore waters caused by the discharging of untreated sewage from coastal towns and cities and from tourist developments
- reclamation of mangroves and coastal wetlands to create land for tourism, housing and agriculture
- the loss and degradation of coral reefs as result of pollution, the 'mining' of coral and sand for building materials, and unsustainable fishing and recreational diving
- the damaging impacts of aquaculture on the marine environment.

In addition to these serious conflicts, Sri Lanka faces two physical challenges.

- Serious coastal erosion along the south-west coast, where many people live and where much of the country's economic wealth is concentrated.
- The exposure of much of the coast to tsunamis and storm surges. The risks associated with this exposure are increasing as a result of global warming. The Indian Ocean tsunami of 2004–05 was a sharp reminder of this hazard (Figure 2.38). More than 30 000 Sri Lankans were killed (either by drowning or by the infectious diseases that followed). Over 1.5 million became homeless. Two important parts of the Sri Lankan economy were badly hit: tourism and agriculture. Coastal roads and railways, particularly in the south-west, were washed away or badly damaged.

Since 1990, Sri Lanka has had a Coastal Zone Management Plan (CZMP). While it recognises all the conflicts and challenges described above, its objectives are rather broad, such as:

- to improve the quality of the coastal environment
- to promote economic development based on coastal resources.

It needs to prioritise and state what needs to be done most urgently, giving specific details about how actions should be taken.

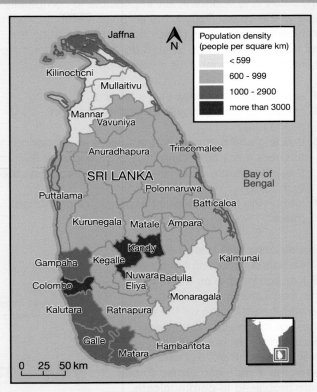

▲ Figure 2.37: The distribution of population in Sri Lanka, 2012

▲ Figure 2.38: The impact of the 2004-05 Tsunami on Sri Lanka

As for the tsunami threat, the CZMP encourages a 'culture of safety'. It recognises that it would be physically and economically impossible to completely defend the coast. Rather, the emphasis is on improving early warning systems (prediction) and educating the public about what they should do during another major tsunami (prevention). This involves:

- promoting awareness of the hazard
- practising emergency drills and evacuation procedures
- building and manning watch towers or anchoring offshore automated warning buoys (Figure 2.39); these buoys are expensive and rely on sophisticated technology
- identifying the areas most exposed to the tsunami threat.

This strategy was tested in 2012. There was a successful evacuation of 1 million people as the result of a tsunami alert created by a powerful earthquake off the west coast of Indonesia. Nationwide alerts were sent out on radio, television and mobile phone networks. Police and armed forces were mobilised to communicate the alert to villages. At the same time, 75 tsunami warning towers set off sirens and monitored the approach of the tsunami. Had this warning system been in place in 2004, the death toll of the Indian Ocean tsunami might have been less.

▶ **Figure 2.39: Tsunami early warning system**

SKILLS ▷ DECISION MAKING

ACTIVITY

Which would you say has a higher priority: dealing with the coastal erosion in south-west Sri Lanka or the pollution of coastal waters? Give your reasons.

SKILLS ▷ REASONING

ACTIVITY

Compare watch towers and automated buoys as tsunami warning systems.

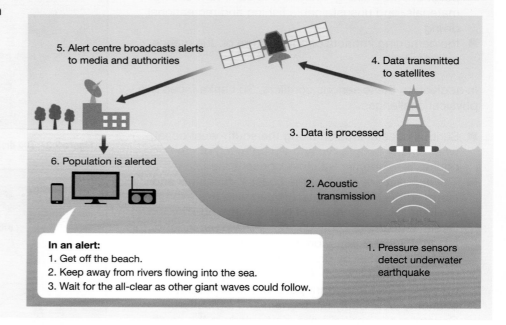

5. Alert centre broadcasts alerts to media and authorities

4. Data transmitted to satellites

3. Data is processed

6. Population is alerted

2. Acoustic transmission

In an alert:
1. Get off the beach.
2. Keep away from rivers flowing into the sea.
3. Wait for the all-clear as other giant waves could follow.

1. Pressure sensors detect underwater earthquake

CASE STUDY: COASTAL MANAGEMENT IN THE UK

Today in the UK and other developed countries, the approach to coastal management has become much more comprehensive. It assesses all the possible risks, such as the risk of flooding and cliff erosion. Those risks are then considered from the perspective of different groups of people (stakeholders) in the coastal community. Who is likely to be most threatened by a particular problem? What do they stand to lose if the particular situation takes place?

A plan of action is then drawn up to minimise the risks and costs. In the UK, the long coastline is divided up into cells (Figure 2.40) and these into smaller sub-cells. A management plan is prepared for each sub-cell. Figure 2.40 shows the management plan for the Isle of Wight. You will see that the coastline is divided up into three different types of action.

- Do nothing – this means that along these stretches there are few if any risks. The coast is safe and secure.

- Hold the line – hard engineering will be needed to continue to protect these stretches of coast because much money is invested in urban development, for example in the tourist resorts of Cowes, Shanklin and Ventnor.
- Managed retreat – three low-lying areas are recognised as being threatened by flooding. Clearly, the retreat at the resort of Bembridge will need very careful management.

The east coast of England is 'sinking' into the sea at a rate of over 6 mm a year. This is beginning to threaten large areas of low-lying coast. Many kilometres of sea wall have been built over the last few centuries as protection. In Essex, about 40 per cent of salt marshes have been lost as a result of rising sea level, coastal erosion and land reclamation. The mudflats and salt marshes here are important feeding and nesting areas for huge numbers of wading birds.

The Royal Society for the Protection of Birds (RSPB) owns Abbotts Hall Farm on the north side of the Blackwater Estuary (Figure 2.42 on page 62). In 2002, it started a programme of managed retreat. Five breaches were made in the old sea walls. This has allowed the sea to cover some 80 hectares of arable fields. The flooded land will gradually revert to what it was before it was cultivated – salt marsh.

Of course, not everyone is a nature lover and supports this or other programmes like it. Many argue that it is a government's responsibility to protect land and people when they are being threatened by the sea. But what about the financial costs, especially in an age of global warming and rising sea levels?

SKILLS INTERPRETATION

ACTIVITY

Study Figure 2.41. What do you notice about locations where 'hold the line' is the coastal management option?

▼ Figure 2.40: The coastal cells of England and Wales

▶ Figure 2.41: Coastal management of the Isle of Wight, southern England

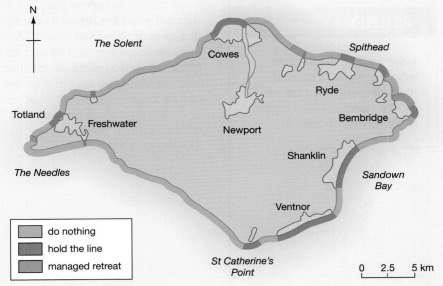

▼ **Figure 2.42: Abbotts Hall Farm managed retreat**

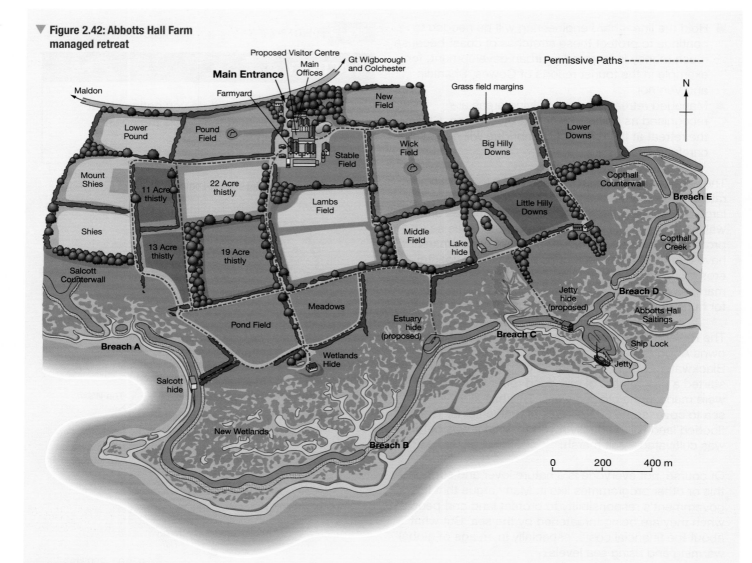

SKILLS CRITICAL THINKING

ACTIVITY

Have a class discussion which compares coastal management in Sri Lanka or The Gambia with coastal management in the UK.

It will take time before people living on the coast all over the world accept that not all of the present coastline can be held against the sea, especially at a time of rising sea levels. Managed retreat at Abbotts Hall Farm was easy because the land being lost was of no great monetary value. Now other authorities need to see that managed retreat is not just good for wildlife. Managed retreat ensures a supply of sediment; this builds up the natural coastal defences of other areas of coastline by forming marshes or beaches, shingle bars or mud flats. Perhaps most important, managed retreat is much cheaper than building sea walls and can leave the coast looking as it did before.

CHAPTER QUESTIONS

END OF CHAPTER CHECKOUT

INTEGRATED SKILLS
During your work on the content of this chapter, you are expected to have practised in class the following:

- using maps to link coastal forms to geology
- using GIS to map coastal systems
- using world maps to show the distribution of coastal ecosystems
- using and interpreting nutrient cycle and food web diagrams
- using maps to show shoreline management plans.

SHORT RESPONSE

1 Identify **one** way in which the geology affects the coast. [1]

2 Identify **one** good and **one** service provided by coastal marshes. [2]

3 State **one** way in which tourism can threaten the coast. [1]

4 Name **two** examples of a coastal stakeholder. [2]

LONGER RESPONSE

5 Suggest **two** conditions required for the formation of sand dunes. [4]

6 Explain the role of longshore drift in the formation of coastal landforms. [4]

7 Analyse how prediction can reduce risk of coastal flooding. [8]

8 Assess the coastal management options for dealing with global warming. [8]

EXAM-STYLE PRACTICE

1 Identify **one** feature formed by coastal deposition. [1]
   A Headland
   B Cuspate foreland
   C Stack
   D Coral reef

2 a) Name **one** type of weathering affecting coastal landscapes. [1]

   b) Name **one** type of mass movement affecting coastal landscapes. [1]

   c) Explain **one** process of coastal erosion. [2]

3 Study Figure 2.14 (on page 41). Explain how raised beaches are formed. [3]

4 Suggest **two** reasons why mangroves are a useful ecosystem. [4]

5 Study Figure 2.17 (on page 43). Identify **one** factor affecting the distribution of coral reefs. [1]

6 For a named country, describe the main conflicts that occur along its coasts. [4]

7 Evaluate soft-engineering and hard-engineering approaches to the management of coastal erosion. [8]

[Exam-style practice total 25 marks]

3 HAZARDOUS ENVIRONMENTS

LEARNING OBJECTIVES

By the end of this chapter, you should know:

- The characteristics of three different types of natural hazard – tropical cyclones, volcanic eruptions and earthquakes

- The causes of tropical cyclones

- The causes of volcanic eruptions and earthquakes

- The scales used in the measurement of different natural hazards

- The primary and secondary impacts of one volcanic eruption and one earthquake

- The reasons why people continue to live in hazardous environments

- That individual tropical cyclones can differ in their impacts

- How people predict and prepare for earthquakes

- Steps in the recovery from a natural hazard.

This chapter is about three different natural hazards that threaten people in many parts of the world. They are tropical cyclones (hurricanes and typhoons), volcanic eruptions and earthquakes. These events have the power to cause great damage to settlements and to injure and kill many people. Is it possible to predict when and where they will occur? What can be done to minimise their destructive impacts, both before and immediately after the event?

3.1 DIFFERENT TYPES OF HAZARD

A **hazard** is an event that threatens, or actually causes damage and destruction to people, their property and settlements. A natural hazard is one produced by environmental processes and involves events such as storms, floods, earthquakes and volcanic eruptions. Some places are more hazardous than others. This is because:

- some places experience more than one type of natural hazard event
- some places experience natural hazards more frequently than others
- in some places the hazards are stronger and more destructive than in others
- some places are better able to cope with the damaging impacts of natural hazards

Table 3.1 classifies hazards into four main categories based on their causes and gives some examples of each. The first three columns relate to natural hazards. These are caused by **natural events** in the environment. The fourth column reminds us that there are also hazards created by people. These range from industrial explosions to nuclear warfare, from air and road crashes to fire and the collapse of buildings. Most of these are the outcome of mishaps to do with human technology. The important point to remember is that if there were no people there would be no natural hazards – just natural events.

▶ Table 3.1: Four major categories of hazard with examples

GEOLOGICAL	CLIMATIC	BIOLOGICAL	TECHNOLOGICAL
Earthquakes	Storms	Fires	Nuclear explosion
Volcanic eruptions	Floods	Pests	Transport accidents
Landslides	Drought	Diseases	Pollution

CHECK YOUR UNDERSTANDING

Are you able to add another hazard to each of the four columns?

Table 3.1 may look neat and tidy, but some hazards may have more than one cause and therefore do not fit easily into this classification. For example, floods are not only caused by the heavy rainfall of a tropical cyclone. A stream of volcanic lava running down into a valley can easily block the flow of a river and cause flooding upstream. Floods in coastal areas can be caused by the tidal waves (tsunamis) associated with earthquakes. Coastal flooding also results from storm surges caused by low atmospheric pressures and not from heavy rainfall (see Part 2.8).

Another point is that some natural events only become hazards in indirect ways. For example, drought becomes a hazard mainly because of its effect on food production. Crops and livestock which lack water will not yield so much food. Food shortages mean malnutrition and possibly death by starvation (Figure 3.1).

Diseases are an interesting group of hazards. Are diseases natural events or are they caused by humans? Some are certainly natural hazards. For example, malaria is a 'vector' or biological hazard carried from one human to the next by a mosquito. In contrast, there are many contagious diseases associated with human pollution of the environment. Typhoid and cholera are just two examples. An interesting aspect of all diseases is that the hazard threat is very focused on people. The outcomes are illness and death for people.

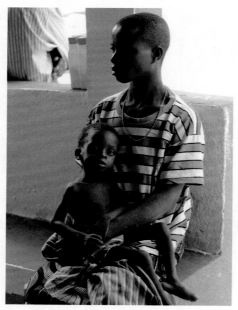

▲ Figure 3.1: Malnourished child in the Sahel, Africa

The focus of this chapter is on three natural hazards: tropical cyclones (storms), volcanic eruptions and earthquakes (Figure 3.2). Tropical cyclones are a climate hazard in which two weather elements, wind and rain, threaten human life and property. Earthquakes and volcanic eruptions are geological or tectonic hazards. (See Parts 1.9 and 2.8 for two hazards – river and coastal flooding – not covered here.)

In Parts 3.2 and 3.3 we will examine three aspects of each type of hazard: their causes, distributions and distinctive characteristics.

▶ **Figure 3.2: The aftermath of an earthquake in Nepal (2015)**

3.2 TROPICAL CYCLONES

CAUSES

A tropical cyclone starts when high temperatures cause air to rise from the surface of the sea. The rising air causes local thunderstorms. Occasionally these small storms come together and create a strong flow of warm, rapidly rising air, which produces an area of increasingly low pressure.

▼ **Figure 3.3: The structure of a tropical cyclone (in the northern hemisphere)**

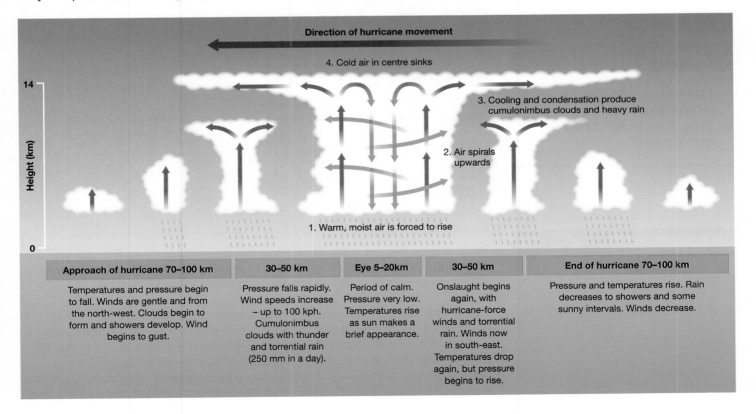

Direction of hurricane movement

4. Cold air in centre sinks

3. Cooling and condensation produce cumulonimbus clouds and heavy rain

2. Air spirals upwards

1. Warm, moist air is forced to rise

Height (km) — 14 / 0

Approach of hurricane 70–100 km	30–50 km	Eye 5–20km	30–50 km	End of hurricane 70–100 km
Temperatures and pressure begin to fall. Winds are gentle and from the north-west. Clouds begin to form and showers develop. Wind begins to gust.	Pressure falls rapidly. Wind speeds increase – up to 100 kph. Cumulonimbus clouds with thunder and torrential rain (250 mm in a day).	Period of calm. Pressure very low. Temperatures rise as sun makes a brief appearance.	Onslaught begins again, with hurricane-force winds and torrential rain. Winds now in south-east. Temperatures drop again, but pressure begins to rise.	Pressure and temperatures rise. Rain decreases to showers and some sunny intervals. Winds decrease.

ACTIVITY

Find out about the Coriolis force. How does it affect a tropical cyclone?

A number of conditions have to be met if these local thunderstorms are to develop into a tropical cyclone:

- a deep layer of humid, warm (>27 °C) and unstable air
- a supply of energy (heat and moisture) from the surface of the sea
- the sea must be at its warmest (during late summer)
- a circulatory motion of the air, anti-clockwise in the northern hemisphere, encouraged by the **Coriolis force**
- small changes in wind speed and direction with increasing altitude (wind shear); this encourages the circulatory motion within the cyclone.

When all these conditions occur, a tropical cyclone is likely to develop very quickly and gather enormous energy and strength. Figure 3.3 shows its basic structure, namely a rotating and vigorous upward spiral of humid, warm air. Curiously, at its very centre (the eye) there is an area of subsiding air with calm conditions and clear skies. Surrounding the eye is the eye wall, where the most destructive energy of the cyclone occurs.

Tropical cyclones are given names by meteorologists. These are from alphabetical lists, with alternating male and female first names over a six-year cycle. There are different lists (and names) for different parts of the world. Names help to identify and track individual cyclones, especially as there may be more than one happening at a time.

DISTRIBUTION

Figure 3.4 shows the distribution of tropical cyclones and what they are called in different parts of the world. For example, in North America, Central America and the North Atlantic they are called hurricanes.

▶ **Figure 3.4: Global distribution of tropical cyclones**

SKILLS INTERPRETATION

ACTIVITY

Look closely at Figure 3.4. What do you notice about the limits to tropical storms in the two hemispheres?

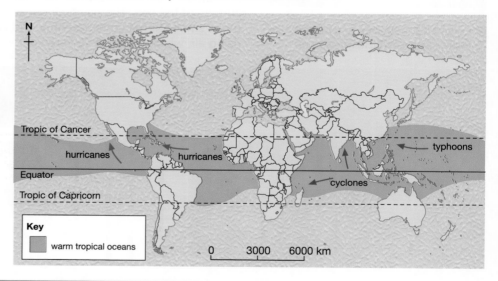

CHARACTERISTICS

Tropical cyclones are areas of very low air pressure (less than 950 mb) and are often 600–700 km across (Figure 3.5). They bring torrential (fast and heavy) rain, thunder and lightning and very strong winds (over 120 km/hr).

Tropical cyclones do not remain where they were formed. They follow the direction of the prevailing winds and ocean currents. The track of a tropical cyclone affects how strong it becomes. The further it travels over the sea, the more energy (heat and moisture) it gathers from contact with warm ocean waters. This increases its strength. When a cyclone reaches land, however, the supply of energy is cut off. It therefore loses strength, moves more slowly and gradually disappears. The average duration of a tropical cyclone is 10 days, but the biggest can last for up to four weeks.

Tropical cyclones can cause widespread damage on land, as well as being a serious hazard to ships at sea. This is caused by:

- the very strong winds – these can destroy trees, crops, buildings, transport links, power supplies and communications
- the torrential rain – this can lead to serious inland flooding and often triggers landslides and mudslides
- the storm surges – these are sudden rises in sea level associated with the very low pressure which allows the sea to 'expand' and the level of the sea to rise (Figure 3.6); storm surges can cause immense damage in coastal areas.

▲ Figure 3.5: A satellite image of Hurricane Katrina over New Orleans (2005). Note the eye at the centre of the hurricane.

▶ Figure 3.6: A storm surge and its destructive power

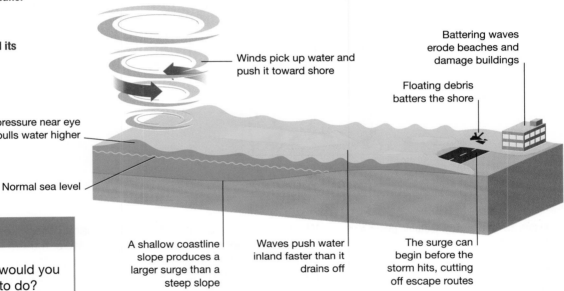

Winds pick up water and push it toward shore

Battering waves erode beaches and damage buildings

Floating debris batters the shore

Low pressure near eye pulls water higher

Normal sea level

A shallow coastline slope produces a larger surge than a steep slope

Waves push water inland faster than it drains off

The surge can begin before the storm hits, cutting off escape routes

SKILLS ▸ REASONING

ACTIVITY

What sort of damage would you expect a storm surge to do?

3.3 VOLCANIC ERUPTIONS AND EARTHQUAKES

Since volcanic eruptions and earthquakes both result from **plate movements**, they are closely linked in terms of both their distributions and causes. They differ, however, in the sort and scale of the damage they do.

CAUSES

The crust of the Earth is made up of a number of tectonic plates, which are the rigid blocks that make up its surface. These plates move over the surface of the globe. Their movements create four different types of plate margin. The **tectonic plates** shown in Figure 3.7 are constantly moving. But they do so at a rate that is almost imperceptible on a human time scale.

When two plates are moving apart, for example in the oceans, the margin between them is called a constructive or divergent plate margin. It is called this because magma (molten rock) rises to the crust to fill the gap and creates new crust through submarine volcanoes. This is happening along the mid-ocean ridge in the Atlantic Ocean.

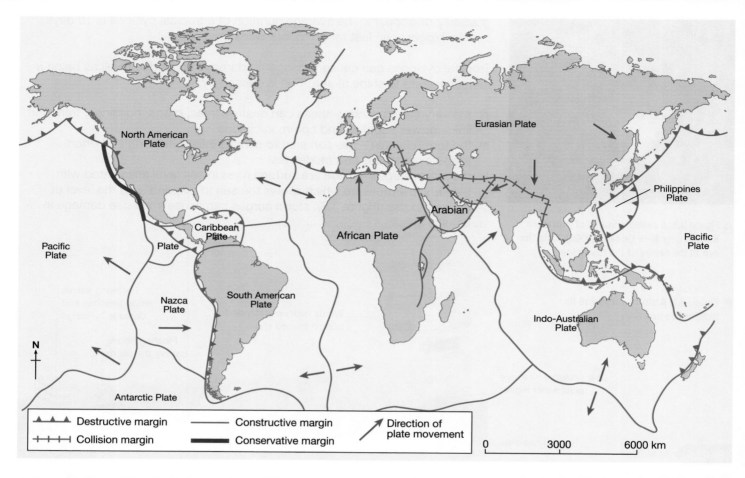

▲ Figure 3.7: The world's tectonic plates and their margins

When two plates are moving towards each other, like the Nazca Plate and the South American Plate, the margin between them is called a destructive or convergent plate margin. The edge of one plate margin is being destroyed as it plunges beneath the other plate that it is meeting head on. This is known as **subduction**. Molten rocks rise to the surface to form volcanoes. The friction between the two plates creates earthquakes.

A collision plate margin occurs where two plates meet head on and are of equal density and strength. The sediments between the two plates are squeezed upwards. The result is the formation of fold mountains, such as the Himalayas, which were created by the collision between the Eurasian Plate and Indo-Australian Plate. Earthquakes are created by the pressure and friction.

There is a fourth type of plate margin referred to as a conservative plate margin. This occurs where two plates are sliding past each other. Since there is neither rising magma here nor subduction, there are no volcanoes. Instead, the friction gives rise to earthquakes as happens in California between the Pacific Plate and North American Plate.

Hotspots are locations beneath the Earth's crust where strong and rising currents of magma (known as plumes) occur. Where the crust above a plume is weak, volcanic activity occurs, as in the Hawaiian islands.

Tectonic plates shape the landscape by creating new rocks and forming mountains and rift valleys. They do this by the processes of volcanic activity, folding and faulting. These are the most powerful natural forces on the planet. It is not surprising that they give rise to the most awesome natural hazards.

DID YOU KNOW?

There are four different types of plate boundary. What sometimes confuses students is that different names are given to the same boundary. This may help to clarify the situation:

- destructive: also referred to as convergent or subduction
- constructive: also referred to as divergent
- conservative: also referred to as transform
- collision: no alternative name.

ACTIVITY

Make a list of the different types of plate margin. Write brief notes about what is happening at each of them.

DISTRIBUTION

SKILLS INTERPRETATION

ACTIVITY

If you compare Figures 3.8 and 3.9, you will see that the African Rift Valley is the location of many volcanoes and few earthquakes. Can you suggest any reasons for this difference?

The distribution of volcanoes and earthquakes is similar in that they both occur along tectonic plate boundaries. A comparison of Figure 3.8 and Figure 3.9 shows how similar the distributions are. There is a concentration of volcanoes along the destructive plate margins that fringe the Pacific Ocean (Figure 3.8). Another concentration is along the African Rift Valley, which coincides with a constructive plate margin. The shores of the Pacific Ocean are also marked by a high density of earthquakes (Figure 3.9). Notice also the occurrence of earthquakes under the Atlantic, Indian and Pacific Oceans. Underwater volcanic eruptions can cause tidal waves (tsunamis) (see page 55).

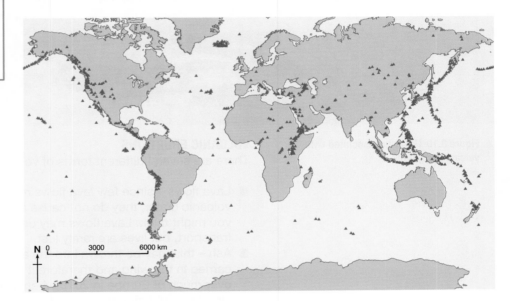

▶ Figure 3.8: Global distribution of volcanoes

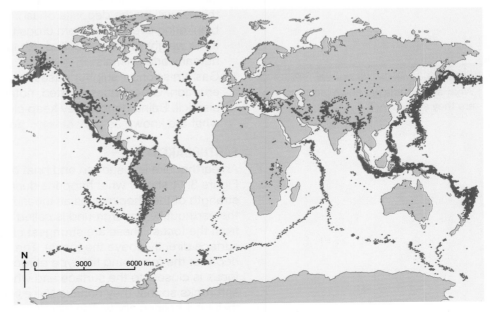

▶ Figure 3.9: Global distribution of earthquakes

Although their distributions and causes are similar, the hazards of volcanic eruptions and earthquakes are very different.

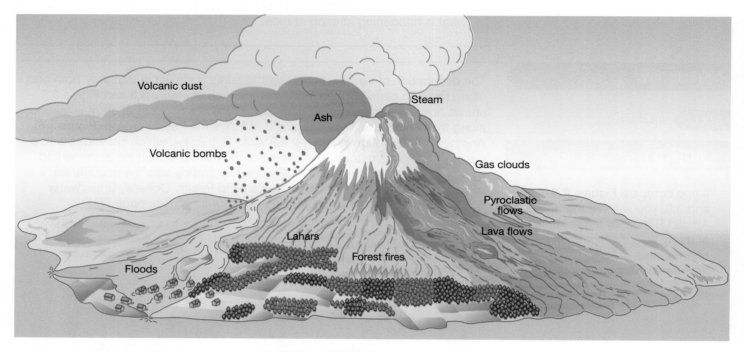

Volcanic dust

Steam

Ash

Volcanic bombs

Gas clouds

Pyroclastic flows

Lava flows

Lahars

Forest fires

Floods

▲ **Figure 3.10: Hazards associated with volcanic eruptions**

VOLCANIC ERUPTIONS

There are several different forms of volcanic eruption (Figure 3.10).

- Lava flows – since few lava flows reach much beyond 10 km from the volcanic crater, they do not cause as much death and destruction as you might think. Lava flows may destroy farmland, buildings and lines of transport, but lives are rarely lost.
- Ash – this may be thrown into the air during a violent eruption. Often ash is carried in the wind and therefore it can affect a large area. This happened over much of Europe in 2010 when a volcano in Iceland erupted. The ash cloud brought air travel to a halt. The further away from the volcano, the thinner will be the deposits of ash. Ash causes much damage by simply blanketing everything, from crops to roads. Roofs of buildings will collapse if the weight of the deposited ash is great. Air thick with ash can asphyxiate humans and animals.
- Gas emissions – sulphur gases are not the only type emitted during an eruption. Other gases emitted, notably carbon dioxide and cyanide, can also kill. Being dense, they keep close to the ground. Particularly lethal are what are known as **pyroclastic flows**.

CHECK YOUR UNDERSTANDING

What are pyroclastic flows, and why are they so deadly?

EARTHQUAKES

An earthquake is a sudden and brief period of intense shaking of the ground. Figure 3.11 shows what happens during an earthquake, depending on the strength of the shockwaves as measured on the Richter Scale. The centre of the earthquake underground is called the focus. Shock waves travel outwards from the focus. These are strongest close to the **epicentre** (the point on the surface directly above the focus). The amount of damage depends on the depth of the focus and the type of rock. The worst damage occurs where the focus is closest to the surface and where rocks are soft. Shock waves 'liquefy' soft rocks so that they behave like a liquid. This means that such rocks lose their load-bearing ability. The foundations of buildings and bridges simply collapse.

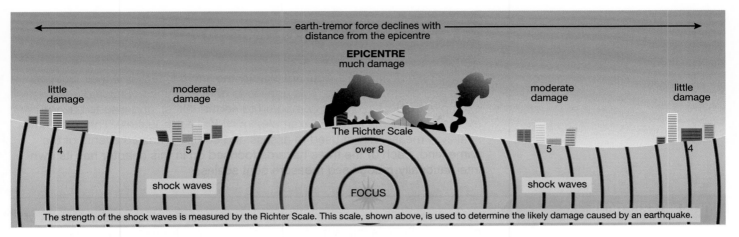

earth-tremor force declines with
distance from the epicentre

EPICENTRE
much damage

little
damage

moderate
damage

moderate
damage

little
damage

The Richter Scale

4

5

over 8

5

4

shock waves

FOCUS

shock waves

The strength of the shock waves is measured by the Richter Scale. This scale, shown above, is used to determine the likely damage caused by an earthquake.

▲ Figure 3.11: A cross-section through an earthquake and the Richter Scale

SKILLS ▷ DECISION MAKING

ACTIVITY

Imagine you are in school and you feel the first tremors of a powerful earthquake. What would you do?

The hazard threat of earthquakes lies in their ability to shake buildings so vigorously that they fall apart and collapse. Often this damage is worse than it should be because of poor building design. There is a saying that 'earthquakes do not kill people, buildings do!'. It is falling masonry that traps, crushes and kills people. Earthquakes rupture gas pipes and break electricity cables. It is not surprising therefore that fire is another aspect of the earthquake hazard. The movement of the ground can be both vertical and horizontal.

1. An underwater earthquake makes the seafloor snap up, lifting a column of water. Gravity pulls the water down, fanning waves outward.

2. Individual waves in a tsunami are spread out; the distance between two wave peaks (the wavelength) can be hundreds of kilometres. Each wave's amplitude (height) is at first rarely not more than 0.9m.

▶ Figure 3.12: How a tsunami forms

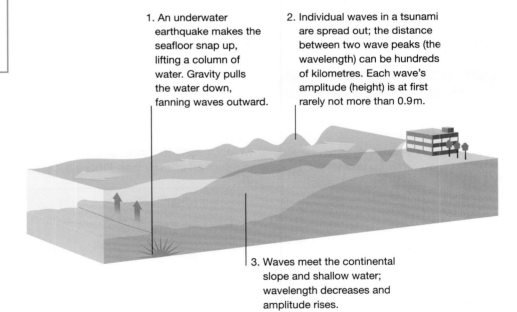

3. Waves meet the continental slope and shallow water; wavelength decreases and amplitude rises.

Another after-effect of an earthquake is the tidal wave or tsunami (Figure 3.12). Earthquakes with epicentres under the sea can generate particularly large and destructive tidal waves. The Indian Ocean tsunami of 2004 had its epicentre just off the west coast of Sumatra and generated a tidal wave up to 30m high. It caused immense damage in the coastal areas of those countries bordering the Indian Ocean. The casualty list amounted to nearly 300 000. The majority of these victims were drowned. It was one of the deadliest **natural disasters** ever recorded.

Volcanic eruptions can also generate tsunamis. The huge eruption of Krakatoa in 1883 created waves up to 35 m high. These waves drowned over 36 000 people.

3.4 THE SCALE OF TECTONIC HAZARDS

No two natural hazard events are exactly the same in terms of their location, scale and destructiveness. This applies to different types of hazard (for example, tropical cyclones compared to earthquakes), and also to hazard events of the same kind (for example, individual earthquakes). There are now scales of measurement used to assess and compare hazard events of the same kind. Each of the three hazards focused on in this chapter has its own internationally-recognised measurement scales.

VOLCANIC ERUPTIONS

Volcanic eruptions can take many different forms. For this reason, the task of measuring them is made much more difficult than, for example, measuring tropical cyclones (see Part 3.7). At present only one measure is fairly widely used, the Volcanic Explosivity Index (VEI).

SKILLS ▶ DECISION MAKING

ACTIVITY

Why do you think it is difficult to measure volcanic eruptions?

The VEI was devised in 1982 and measures the intensity of volcanic eruptions on a logarithmic scale of 1 to 8. In this respect, it is similar to the scales used to measure earthquakes (see below). Explosivity is measured in terms of the volume of ash (tephra) ejected and the height that ash reaches into the atmosphere. The limitations of these measures are obvious, if only because ash is not the main product of all hazardous volcanic eruptions. While some eruptions are explosive with rock and ash blasted from the vent, others are effusive with lava flowing from the vent. Therefore, the VEI is only suited to the assessment of one type of volcanic eruption.

EARTHQUAKES

We use three different scales to measure the strength of earthquakes. The Richter Scale measures an earthquake's strength according to the amount of energy released during the event. The energy is measured by a seismograph. The Richter Scale runs from 2.4 or less to over 8.0. It is a logarithmic scale, which means that one point up on the scale represents a 30-times increase in released energy (see Figure 3.11 on page 73).

▼ Figure 3.13: Some examples from the Mercalli Scale

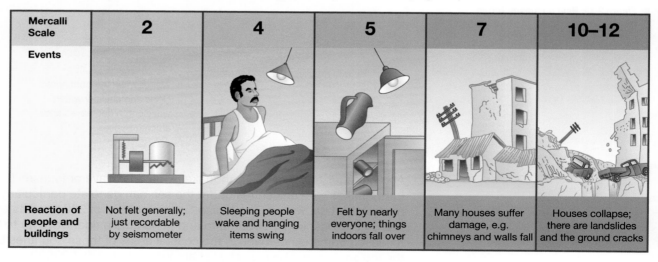

Mercalli Scale	2	4	5	7	10–12
Events					
Reaction of people and buildings	Not felt generally; just recordable by seismometer	Sleeping people wake and hanging items swing	Felt by nearly everyone; things indoors fall over	Many houses suffer damage, e.g. chimneys and walls fall	Houses collapse; there are landslides and the ground cracks

The Mercalli Scale is quite different. It is based on what people experience and the amount of damage done (Figure 3.13).

The Moment Magnitude Scale (MMS) was introduced in 1979 and is now the most commonly used method of measuring earthquakes. Like the Richter Scale,

ACTIVITY

Which of the three earthquake measures do you prefer? Give your reasons.

ACTIVITY

Look back at Figure 3.11. How and why do Richter Scale values vary across an earthquake?

ACTIVITY

Explain why capital and technology are so important when dealing with natural hazards.

it measures the energy released by an earthquake on a logarithmic scale. But the assessment of energy is based on the amount of rock movement along a fault or fracture, and the area of the fault or fracture surface. Scientists think that this is a more accurate way of measuring and comparing earthquakes.

The amount of damage and destruction caused by a particular natural disaster does not just depend on its scale and destructive energy. Others factors are involved, as listed below.

- The size of the area affected.
- The density of population in the area affected; the more people and economic activities in a disaster area, the greater the potential damage.
- How long the event lasts.
- The degree to which people are warned in advance of the event. This is one reason why earthquakes are often so devastating. They occur almost anywhere near a plate margin without warning.
- The degree to which people are prepared for a possible natural hazard. Are there emergency shelters? Have people been educated in what should be done in an emergency? Are houses, factories and businesses located in areas of low risk? Have buildings been constructed so that they can withstand the hazard?
- The ability of a country or region to cope with the aftermath of a hazard, both immediately and in the longer term; a lack of capital and technology explains why natural hazards often have a worse impact on developing countries.

There is a significant difference between developing and developed countries and their ability to prepare for hazards and to cope with the damage. This important point is explored further with respect to tropical cyclones in Parts 3.7, 3.8 and 3.9.

3.5 IMPACTS OF TECTONIC HAZARDS

In this part, we take a closer look at the impacts of volcanic eruptions and earthquakes through two contrasting case studies of each hazard.

In hazard studies today, we distinguish between a hazard's primary impacts (its immediate effects) and secondary impacts (effects that happen later on). The primary impacts of an earthquake are a direct result of the ground being shaken, for example:

- collapsed buildings and the people killed or injured by falling masonry
- broken water, gas and sewage pipelines
- downed electric power lines.

The secondary impacts include:

- tsunamis
- aftershocks
- fires due to ruptured gas mains.

In the case of volcanic eruptions, the primary impacts include:

- buildings, roads and crops destroyed by lava flows
- death and injuries resulting directly from the outpourings of lava, ash and gas
- contamination of water supplies by ash falling on them.

The secondary impacts include:

■ **lahars** created by the mixing of volcanic ash and mud with rainwater or melting snow
■ fires started by lava and pyroclastic flows
■ the psychological trauma of losing family members and friends; this is a secondary impact of all types of hazard event.

Let us now consider two earthquake events, one in Nepal and the other in Italy.

CASE STUDY: EARTHQUAKES IN NEPAL (2015)

On April 25, 2015, a massive earthquake (magnitude 7.8 on the Richter Scale) devastated much of the small Himalayan country of Nepal (Figure 3.14). It caused over 8500 deaths and nearly 17 000 people were injured. There was significant damage to homes and buildings such as shops, medical centres and schools. Nearly 3 million people were displaced from their homes.

Other primary impacts were related to the damaged **infrastructure**. Thousands of families instantly lost access to household energy for cooking, lighting and heating. Without power to charge mobile phones, they could not contact family members and friends or call

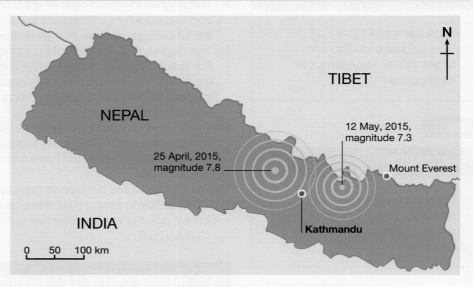

▲ Figure 3.14: The locations and impacts of Nepal's major earthquakes (2015)

for help. The road network which provides the only means of access to many parts of Nepal was badly damaged. A major primary impact of the earthquake was landslides triggered in the mountainous valleys. These removed or blocked long stretches of road. This made it impossible for emergency services to reach many communities that had been badly hit by the earthquake. The smallness of the airports also slowed down the emergency response. This meant that large planes carrying emergency teams and equipment were unable to land. Therefore, the global efforts to help the Nepalese people were frustrated at the start.

Three weeks later, another major earthquake occurred. Its epicentre was a short distance to the east of the April event (Figure 3.14). The aftershock caused mass panic as many people were living in the open after the April earthquake. Perhaps because of this, casualties were much lower, with 155 people killed and 3200 injured.

The Nepalese are slowly coming to terms with this double earthquake and gradually repairing the damage. They are a resilient people; earthquakes are a feature of their everyday life. They have no option other than to live with the earthquake threat. To help the situation, there has been a large inflow of international aid.

However, there is perhaps one thing for which the Nepalese should be grateful. Nepal is a land-locked country tucked well away from the sea. Therefore, it was spared the most notorious of an earthquake's secondary impacts – a tsunami!

SKILLS ▶ INTERPRETATION

ACTIVITY

Locate Nepal on Figure 3.7 (page 70). Along what sort of plate boundary did these earthquakes take place?

CASE STUDY: EARTHQUAKES IN CENTRAL ITALY (2016)

Earthquakes are nothing new in central Italy (Figure 3.15). Here the Apennine mountains have been formed by the Adriatic Plate being subducted beneath the Eurasian Plate. Between August and early November, 2016, however, there was a sudden flurry of earthquakes (Table 3.2).

DATE	MAGNITUDE	EPICENTRE LOCATION	FOCUS DEPTH
24 August	6.2	Amatrice	4 km
26 October	6.1	Visso	8 km
30 October	6.6	Norcia	4.5 km
3 November	4.8	Pieve Torina	8 km

▲ Table 3.2: Earthquakes in central Italy (2016)

The August earthquake did considerable damage to buildings. Around 300 people were killed and over 350 people injured, mainly in the town of Amatrice. The emergency services rescued about 250 people from the rubble of collapsed buildings.

Although three of the earthquakes were roughly of similar magnitude, the last three caused much less damage than the first. People were injured but no one was killed. Possibly this was because the August event put people in this region on the alert. The 30 October earthquake was the most powerful earthquake to strike Italy in 36 years.

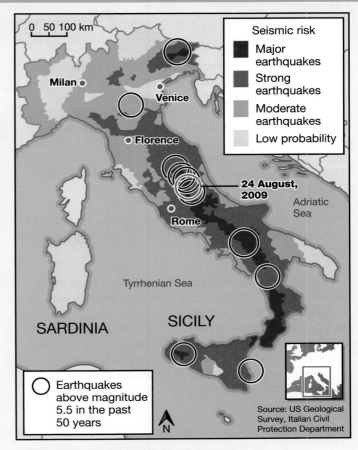

▲ Figure 3.15: The Apennines – a quake-prone region

SKILLS DECISION MAKING

ACTIVITY

Discuss, in class, the possible reasons why the Nepal earthquakes were more destructive than the Italian ones.

This case study illustrates two important points about earthquakes as shown below.

- Even developed countries cannot make themselves completely quake-proof. Many of the damaged buildings in Amatrice dated from before the introduction of quake-proofing building regulations. Also, many new and renovated buildings had ignored those regulations.
- The damage and destruction caused by an earthquake depends not only on its magnitude and its focus depth. There is a major chance factor associated with the epicentre's location. If an epicentre is immediately beneath a town or city, then potential damage could be considerable. If it is located beneath a sparsely populated rural area, then damage will be limited.

SKILLS INTERPRETATION (GRAPHICAL)

ACTIVITY

Plot the four earthquakes in Table 3.2 on a graph. Can we see any relationship between magnitude and focus depth?

SKILLS DECISION MAKING

ACTIVITY

Why do shallow-focus earthquakes tend to be more destructive than deep ones?

We will now look at two volcanic eruption case studies, one in Indonesia (an emerging country) and the other in Japan.

CASE STUDY: MOUNT MERAPI ERUPTION (2010)

Mount Merapi is located on the densely populated island of Java, Indonesia. It rises to nearly 3000 m above sea level and has been erupting frequently for over 600 years. It is one of the most active and dangerous volcanoes in the world. The volcano occurs on the destructive boundary where the Indo-Australian Plate is being subducted beneath the Eurasian Plate. This volcano is one of many active volcanoes in the Pacific Ring of Fire.

The main phase of the eruption took place between 26 October and 12 November, 2010. The primary impacts were:

- volcanic bombs – which landed over 11 km away
- ash plumes blasted up to 5 km into the sky with ash landing up to 30 km away
- pyroclastic flows spreading 3 km down the volcano
- sulphur dioxide, a gas that is dangerous to human health, was blown as far south as Australia
- lahars – ash, rock and lava deposited on the sides of the volcano was washed down into local towns by rainfall.

The secondary impacts of the eruption were quite varied:

- because of the ash and gas, emergency shelters could not be located closer than 15 km from the volcano
- the danger area extended 20 km from the volcano and as a result nearly 300 000 people had to leave their homes
- flights in Western Australia were grounded because of the dangerous ash clouds
- damage to crops resulted in local food shortage.

Source: www.miavita.brgm.fr

▲ Figure 3.16: A risk map of the Mount Merapi area one year after the eruption

ACTIVITY

Locate Mount Merapi on Figure 3.7 (page 70).

Figure 3.16 shows that a year after the eruption there were large areas immediately to the south and west of the vent where levels of risk (medium and high) meant that local people were not allowed to return to their homes. Figure 3.17 shows how much the volcanic cone was split by the eruption. The one positive note is that human costs of the eruption were relatively small in such a densely populated area. There were about 350 deaths, many as a result of pyroclastic flows. Many people suffered burns and breathing problems, but the exact number is not known.

▼ Figure 3.17: The profile of Mount Merapi following the eruptions in 2010

CASE STUDY: MOUNT ONTAKE ERUPTION (2014)

On 27 September, 2014, Mount Ontake, an active volcano and the second highest volcano (3067 m) in Japan, suddenly erupted without any warning. Around 50 hikers, out enjoying a day in the mountains, were killed by the toxic gases and heat of a large and fast-moving pyroclastic flow (Figure 3.18). Nearly 100 people were injured. Emergency services did, however, manage to rescue a number of stranded people.

The impact of the eruption would have been much worse had this part of the island of Honshu been populated. In fact, nearly three-quarters of Japan is mountainous and unpopulated. Nearly everyone lives on fragmented lowlands strung out along the coast. The mountainous interior offers wilderness and recreational space where people can escape the pressures of urban living, if only for a day or weekend. In much of the mountainous interior, the only settlement takes the form of tourist cabins and hostelries.

The obvious question after the event was: why was there no warning of this eruption? After all, Japan has 110 active volcanoes and a very sophisticated volcano-monitoring system. But there were none of usual warning signs such as earth movements or changes

▲ Figure 3.18: One of the few hikers lucky to survive the pyroclastic flow

in the mountain's surface. Scientists point out that this was, in fact, a relatively small eruption. It was caused by the build-up of superheated steam and ash rather than of lava in the magma chamber.

This case study illustrates that even in the 21st century, and with all the latest technology, some tectonic events remain unpredictable. They can still be a nasty surprise.

SKILLS REASONING

ACTIVITY

Suggest reasons why the Mount Merapi eruption caused more deaths and damage than the Mount Ontake eruption.

ACTIVITY

Illustrate the difference between the primary and secondary impacts of a natural hazard.

SKILLS CRITICAL THINKING

ACTIVITY

Thinking back to Chapters 1 and 2, do floods have both primary and secondary impacts? If so, provide some examples.

3.6 REASONS FOR LIVING IN HIGH-RISK AREAS

SKILLS INTERPRETATION

ACTIVITY

Compare Figures 3.4 and 3.9 with Figure 3.19.

▼ **Figure 3.19: Global distribution of 'million' cities**

History tells us where in the world specific types of natural hazard are likely to occur. We have a fairly good idea of where the high-risk areas are. Figures 3.4 (page 68) and 3.9 (page 71) show the global distributions of two natural hazards (tropical cyclones and earthquakes) that cause the greatest number of deaths and the most damage. Compare those maps with the map showing the global distribution of cities with populations greater than one million and therefore areas with high population densities (Figure 3.19). It has been estimated that between 2000 and 2010 about 1 million people were killed by earthquakes. This is hardly surprising because there are billions of people living in the world's earthquake zones. Many of the cities shown in Figure 3.19 are located within the risk areas of earthquakes and tropical cyclones.
So, why do so many people continue to live and work in what are clearly hazardous areas?

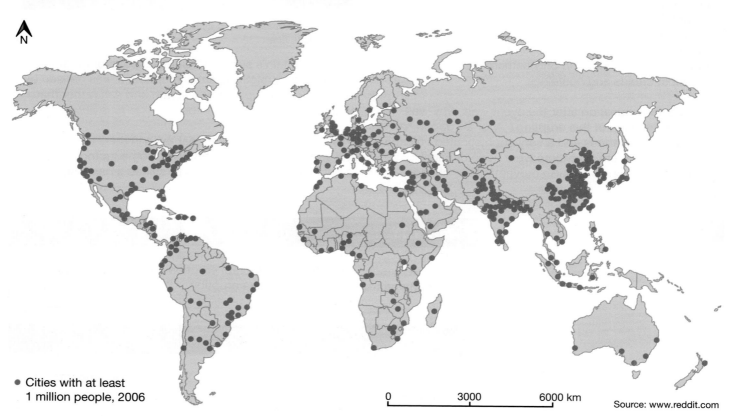

• Cities with at least 1 million people, 2006

0   3000   6000 km

Source: www.reddit.com

There are a number of possible explanations.

■ A lack of education and information may mean that residents are unaware of the real risks, particularly if the hazards occur infrequently. This can be the case particularly in poor undeveloped areas.
■ People may be aware of the risks but decide to live in the area anyway. Perhaps the area offers some attractive benefits (see next page).
■ People may be unable to move away from hazardous areas, owing to a lack of money, or they worry about not being able to find a job elsewhere.
■ Many people are optimists and think that they will never be a natural disaster victim. Alternatively, they may be resigned to their fate – if they are to be victims, there is nothing they can do about it.

▲ Figure 3.20: Fertile soil is one benefit of living near a volcano.

SKILLS ⟩ REASONING

ACTIVITY

Are you able to suggest any more reasons why people continue to live in high-risk areas?

SKILLS ⟩ CRITICAL THINKING

ACTIVITY

In groups, discuss why you think tourists like to visit volcanoes.

■ Perhaps the biggest factor is that the areas of high population density have gradually grown up over many centuries. As a result, they have a momentum which keeps them going no matter how many hazards occur. So, for millions of people, these high-risk areas are home and as a result have a number of attractions. For example, this is where their family has lived for generations. This is where many of their relatives and friends are living today. This is where they work. This is where, despite the hazards, they feel comfortable.

■ The cities in these high-risk areas represent centuries of investment – of money and human effort. No human society is so rich that it can afford to throw away all this investment and abandon these cities.

Unlike earthquakes and tropical storms, which are probably the most vicious and costly natural hazards, volcanoes do offer some benefits (Figure 3.20).

■ Minerals – volcanoes bring valuable mineral resources to the surface. These include diamonds, gold and copper.

■ Fertile soils – volcanic ash often contains minerals that enrich the soil. Fine dust is quickly mixed into the soil and acts as a fertiliser. However, not all volcanoes give rise to fertile soils.

■ Geothermal energy – water running through the Earth's crust is heated by volcanic rock at or near a plate margin. This hot water emerges as hot springs and can be used to heat homes, factories and business premises.

■ Tourism – volcanoes interest many people and attract tourists. Mount Vesuvius (Italy) is a classic example, drawing hundreds of thousands of tourists each year (Figure 3.21). The hot springs in volcanic areas around the world also attract visitors.

▶ Figure 3.21: Mount Vesuvius crater, Italy, is a tourist attraction.

Finally, dense populations are also found in the high-risk areas of other natural hazards. Obvious examples are river valleys and deltas that suffer from regular and severe flooding; for example, the Ganges-Brahmaputra-Meghna delta where 90 million people live. Here, as with volcanoes, there are some benefits such as fertile soils replenished by the regular flooding.

3.7 TROPICAL CYCLONES AND THEIR IMPACTS

MEASUREMENT

CHECK YOUR UNDERSTANDING

What are the differences between Category 2 and Category 4 tropical cyclones?

▼ Figure 3.22: The Saffir-Simpson classification of tropical cyclones

Of the three hazards examined in this chapter, the tropical cyclone is the most easily measured. Rainfall and wind speeds, two key aspects of tropical cyclones, are constantly monitored around the world by satellites and a network of weather stations. This 24-hour availability of weather data also gives us an advantage over tropical cyclones that we do not have over volcanic eruptions and earthquakes. Their existence and approach can be forecast with some accuracy. Advanced warnings can be given.

Once wind speeds in a tropical storm reach 119 kph, it is classified as a cyclone and the Saffir-Simpson Scale (Figure 3.22) is used to assess it. The scale recognises five categories of cyclone strength based on four features: wind speed, air pressure, storm surges and typical damage.

Saffir-Simpson hurricane scale			
category	winds (kmph)	damage	storm surge (m)
1	74–95	**Minimal:** Damage to unanchored mobile homes,vegetation and signs. Coastal road flooding. Some shallow flooding of susceptible homes.	4–5
2	96–110	**Moderate:** Significant damage to mobile homes and trees. Significant flooding of roads near the coast and bay.	6–8
3	111–130	**Extensive:** Structural damage to smallbuildings. Large trees down. Mobile homes largely destroyed. Widespread flooding near the coast and bay.	9–12
4	131–155	**Extreme:** Most trees blown down. Structural damage to many buildings. Roof failure on small structures. Flooding extends far inland. Major damage to structures near shore.	13–18
5	More than 155	**Catastrophic:** All trees blown down. Some complete building failures. Widespread roof failures. Flood damage to lower floors less than 15 feet above sea level.	Greater than 18

IMPACTS

The short-term impacts of tropical cyclones fall under four major headings:

- physical – the damage to property caused by high winds, heavy rainfall and storm surges along coasts
- social – the number of people killed or injured; the disruption of communities and their normal way of life; any decline in the quality of life
- economic – the disruption and destruction of businesses, transport links and services
- environmental – landslides; soil erosion; upset ecosystems; prolonged flooding.

As with tectonic hazards, there is a difference between the primary and secondary impacts of a tropical cyclone. The primary impacts are the direct and immediate results of high winds, torrential rain and storm surges. For example, flooding and wind can damage buildings. The secondary impacts are wide ranging, from the costs of repairing the damage to the spread of waterborne disease; and from the loss of homes and personal possessions, to transport lines being blocked by landslides.

Tropical cyclones vary in strength and destructive power, and are not evenly distributed. Some countries are hit more than others (Table 3.3).

▶ Table 3.3: Ranking of countries by tropical cyclone hits since 1970

RANK	COUNTRY	RANK	COUNTRY
1	China	6	Australia
2	Philippines	7	Taiwan
3	Japan	8	Vietnam
4	Mexico	9	Madagascar
5	USA	10	Cuba

SKILLS PROBLEM SOLVING

ACTIVITY

Which continent is not represented in Table 3.3? Which continent is most represented?

Some countries are better able to protect themselves from the threat of tropical cyclones. Figure 3.23 compares the casualty figures of tropical cyclones hitting countries in Asia and the Pacific region. The information is dated but is valid when making comparisons. It is interesting to compare the relatively small number of casualties in Japan with those in China and the Philippines. All three countries have received more than 100 typhoon hits since 1970. The casualties were significantly less in Japan even when taking into account differences in population. Was this because Japan is a developed country?

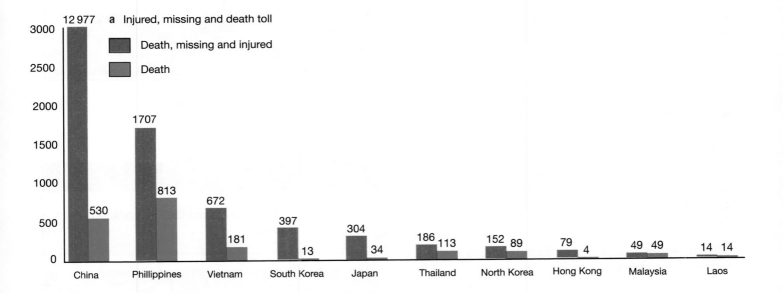

a Injured, missing and death toll

▶ Figure 3.23: Tropical cyclone losses in countries of Asia and the Pacific Region: a) people, b) economic

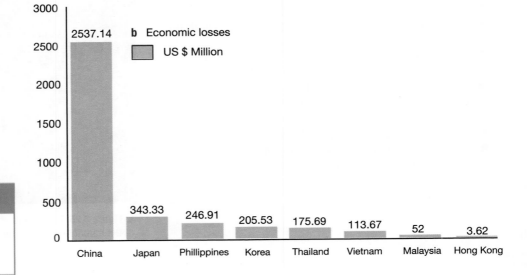

b Economic losses

SKILLS REASONING

ACTIVITY

What would be the 'economic losses' shown in Figure 3.23b?

CASE STUDY: HURRICANE KATRINA HITS THE USA (2005)

Hurricane Katrina formed over The Bahamas in August, 2005, and then moved westwards across the tip of Florida as a Category 1 hurricane (Figure 3.24). It reached Category 5 at sea on 25 August (see Figure 3.22 on page 82). The cyclone doubled in size and wind speeds were over 300 kph. Eventually, it swung north-eastwards, weakened to Category 3 and made its landfall at New Orleans at 6 a.m. on 29 August.

Hurricane Katrina was one of the deadliest and costliest hurricanes in history. Nearly 2000 people were killed, thousands were injured, and damage was estimated to be $84 billion. It displaced 1 million people.

The high winds destroyed many buildings in downtown New Orleans, including hotels and the roof of the city's famous Superdome (Figure 3.25). The hurricane produced a storm surge with 8 m waves which overwhelmed the city's flood protection system. The inrush of sea water broke through the protective levees along the Mississippi Gulf outlet. This, combined with very heavy rainfall, resulted in 80 per cent of the city being flooded up to depths of 6 m (Figure 3.26).

International news coverage of the hurricane and its aftermath prompted people to question why a leading city in the USA should suffer so much devastation and chaos during the days immediately following the event.

Subsequent investigations criticised aspects of the hazard management:

- the levee and coastal flood protection walls were ageing and neglected
- the slow emergency response by the authorities
- the relief efforts of a large number of organisations were poorly coordinated
- food and water shortages were evidence that emergency aid was poorly organised
- despite large-scale evacuations, many people were left behind and many refused to leave their homes
- there was much looting of abandoned properties
- some of the money raised by public appeals around the world 'disappeared' so could not be put to good use.

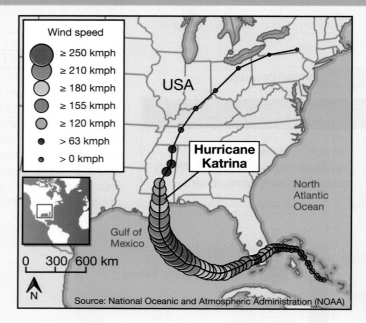

▲ Figure 3.24: The track of Hurricane Katrina, USA (2005)

SKILLS INTERPRETATION

ACTIVITY

Looking at Figure 3.24, describe the track of Hurricane Katrina.

▲ Figure 3.25: Thousands of New Orleans residents were housed in the damaged Superdome.

More than 10 years after the disaster, the city has recovered and flood defences have been improved. Damaged buildings have been repaired or replaced. However, there is much concern about what has happened to the people who were living in some of the worst of the flooded areas. The 'new' New Orleans is now claimed to be more expensive and gentrified. In other words, the once-poor parts of the city have become gentrified. The city may be losing some of its traditional character and culture.

It is hoped that lessons have been learned and that the management of any future hurricane hazard in and around New Orleans will be more effective.

SKILLS ▶ DECISION MAKING

ACTIVITY

Discuss, in groups, which of the criticisms of disaster management was the most serious.

▲ Figure 3.26: A flooded suburb of New Orleans after Hurricane Katrina

CASE STUDY: TYPHOON HAIYAN HITS THE PHILIPPINES (2013)

On 8 November, 2013, a Category 5 typhoon struck the Philippines. Typhoon Haiyan (named in the Philippines as Typhoon Yolanda) originated in the north-west Pacific Ocean. It was one of the most powerful tropical cyclones to hit the Philippines. Winds speeds of over 300 kph were recorded.

As the typhoon approached the Philippines, and made its first landfall, the trail of destruction began. Approximately 800 000 people were evacuated and 1.9 million were left homeless. More than 6000 people were killed and many more injured. Remember, however, that the Philippines is a densely populated country. Quick action by the authorities, aid agencies and local organisations helped to save many lives. But the delivery of emergency aid was very difficult because the airport was badly damaged and roads were blocked by fallen trees

▲ Figure 3.27: The aftermath of Typhoon Haiyan

and debris (Figure 3.27). Much coastal shipping was destroyed by the storm surge generated by the typhoon (Figure 3.28).

▲ Figure 3.28: Storm surge damage caused by Haiyan

▲ Figure 3.29: The delivery of relief

With crops wiped out, there were serious shortages of food. At one time, around 3 million people were receiving food assistance including rice, high-energy biscuits and emergency items. Much of the food was supplied by foreign governments and international relief organisations (Figure 3.29).

Management of this hazard event has now moved to the recovery stage (see Figure 3.36 on page 91). The slogan adopted by the government of the Philippines for the recovery programme is 'Build Back Better'. The aim is that all rebuilt structures will be upgraded and made more hazard-proof. 'No-build' zones have been defined along those stretches of the coast that were hit by the storm surge. A new storm surge warning system is now operational. Mangroves have been replanted to help the natural protection of the coast.

Given that the Philippines is not a rich or technologically advanced country, the management of Typhoon Haiyan, its immediate aftermath and the subsequent recovery, reflects well on the country. It has learned useful lessons from the typhoons that hit them every year. As a result, it has simple but effective forms of hazard preparation (see page 87). These include, for example, building reliable warning systems and holding regular emergency training exercises in local communities. Knowing what to do when a typhoon hits can and does save lives.

SKILLS ANALYSIS

ACTIVITY

Discuss, in groups, which of the recovery options in the Philippines would be: a) the cheapest, and b) the most effective.

ACTIVITY

Both tropical cyclones (Hurricane Katrina and Typhoon Haiyan) were at one time graded Category 5 cyclones. What does this mean?

It is widely believed that developed countries can cope with the impacts of natural hazards better than developing countries. Better access to resources and technology means better hazard preparation, better emergency services and a quicker recovery. A comparison of the previous two case studies suggests that this might not be true. Remember that both cyclones were Category 5. So there must be there other factors underlying the differences between the two events. For example:

■ The size of the population exposed to the hazard. The cluster of islands that make up the Philippines had a population of 100 million in 2013. The population of the New Orleans metropolitan area was just over 1 million. Population densities in the Philippines average 890 persons per square km, while in New Orleans it is only 343. So, although there were three times as

SKILLS ▶ REASONING

ACTIVITY

Write a short account explaining what is involved in hazard management.

many deaths in the Philippines, relatively speaking the death rate was lower than in New Orleans.

- Warning time – Katrina was expected to travel westwards to Mexico, but it suddenly turned north. Within less than 24 hours it had come ashore at New Orleans. Haiyan, in contrast, maintained a fairly consistent direction across the Pacific, and people in the Philippines had almost five days in which to prepare. This was critical in that there was sufficient time to evacuate people from the most exposed areas and to organise emergency services. In New Orleans, the short warning encouraged panic and chaos rather than good hazard management.

With both cyclones, it was the elderly and poor people who suffered most. In terms of recovery, clearly the time difference between the two hazard events means that New Orleans is 8 years ahead on the recovery path. In New Orleans, much of the costs of recovery has been met by insurance. In the Philippines, insurance is a luxury which few can afford. The result is that the poor have struggled just to survive; recovery is still a long way off.

These two case studies suggest that coping successfully with the aftermath of a tropical cyclone (or any natural hazard) depends on the quality of hazard management. Having a fair amount of warning time and a management strategy that suits local circumstances are key. So, too, are effective managers who direct relief operations.

3.8 PREDICTING AND PREPARING FOR EARTHQUAKES

▶ Figure 3.30: Steps in the management of natural hazards

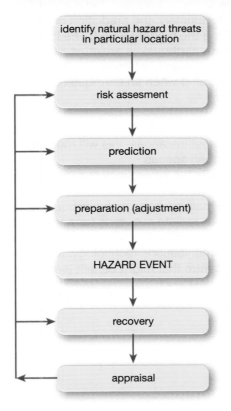

The obvious way of reducing the destructive impact of a hazard event is:

- to know when and where the hazard will occur (prediction)
- to take precautions before the event actually occurs (preparation).

Figure 3.30 outlines these two basic actions as a sequence of steps for managing most natural hazards, not just earthquakes. In this section and the next, it is important that you refer back to the two earthquake case studies on pages 76 and 77.

RISK ASSESSMENT

A very important first step is **risk assessment**. Risk is about the probability of a particular event happening and the scale of its possible damage. Risk is also what people take knowing that they are 'exposed' to a natural event that might prove hazardous. The greater the probability of a natural event occurring in a particular location and causing damage, the greater the risk that people take. This is particularly so if they remain in that location or do not take preventative (evasive) or precautionary actions.

In order to assess risk, we need to be aware of three features of a particular type of hazard, as follows.

■ Its distribution – where does it generally occur on the Earth's surface? Figure 3.9 on page 71 shows where earthquakes occur.
■ Its frequency – how often does it occur? This is tricky in the case of earthquakes because the timing of their occurrence seems quite random.
■ Its scale – does the natural event vary in its degree of hazard and impacts? See the Mercalli Scale and the Moment Magnitude Scale on page 74.
■ Its predictability – does the hazard always behave in the same way? While tropical cyclones show a recognised sequence of events as they pass (see Figure 3.3 on page 67), earthquakes and volcanic eruptions are much more individual.

Another important issue in any risk assessment of earthquakes is location. For example, how exposed is it to the coast and possible tsunamis? Remember that many earthquakes occur out at sea and it is the resulting tsunamis that cause much of the damage and destruction.

SKILLS EXECUTIVE FUNCTION

ACTIVITY

Draw a diagram to show what is meant by the term 'risk assessment'.

CHECK YOUR UNDERSTANDING

Check that you understand why the destructive energy of an earthquake is often most threatening at the coast.

PREDICTION

Prediction is knowing that a hazard event is shortly about to take place. Unfortunately, the prediction of earthquakes is still one of the great challenges facing the modern world. The theory of plate tectonics is helpful in predicting where earthquakes are most likely to occur, but scientists are still unable to predict exactly when they will occur. What is needed is a system that can give a warning when an earthquake is about to happen. In other words, a system that:

■ gives people time to move into what are thought to be 'safe' locations. In most instances, this will mean leaving all buildings and moving out into open spaces
■ puts the emergency services on immediate alert

The main problem with earthquakes is the speed with which the shock waves radiate outwards from the epicentre. Once an earthquake has started, warnings of only a matter of minutes might be possible, even with modern communications technology. It is only those areas more distant from the epicentre that might benefit (see warning systems on page 91).

PREPARATION

▶ **Table 3.4: The deadliest earthquakes (2005-2015)**

SKILLS INTERPRETATION

ACTIVITY

Plot the data in Table 3.4 for magnitude and deaths on a graph. Are you able to detect any trends in your completed graph?

DATE	LOCATION	MAGNITUDE	DEATHS
April, 2015	Nepal	7.8	8,500
August, 2014	China	6.2	700
September, 2013	Pakistan	7.7	825
March, 2011	Japan	9.0	20,000
February, 2010	Chile	8.8	700
January, 2010	Haiti	7.0	316,000
September, 2009	Indonesia	7.5	1,100
May, 2008	Chile	7.9	90,000
August, 2007	Peru	8.0	500
May, 2006	Indonesia	6.3	5,700
October, 2005	Pakistan	7.6	80,000
March, 2005	Indonesia	8.6	1,300

CHECK YOUR UNDERSTANDING

Which is potentially the more damaging: a shallow- or a deep-focus earthquake? Give your reasons.

Table 3.4 makes the important point that the deadliest earthquakes are not necessarily those with the greatest magnitude. Other key factors include population density in the affected area and the degree of earthquake preparation undertaken. The 2010 Haiti earthquake was responsible for many fatalities because it occurred in a developing country with high population densities and a low level of earthquake preparation. By contrast, the much stronger 2011 Japan earthquake (the most powerful ever recorded) was less deadly because this highly-populated country is one of the best prepared in the world for earthquakes and tsunamis. It has to be, as a country located in one of the most tectonically active parts of the world (see Figure 3.9 on page 71). In Japan alone there are as many as 1500 earthquakes recorded every year, many of them with a magnitude of 4.0 or more.

The death toll of the 2015 Nepal earthquake (see Table 3.4 and page 76) might have been higher had the earthquake not mainly struck sparsely populated rural areas. The same applies to the 2016 earthquakes in Central Italy (see page 77). Even so, there was criticism in both locations that the level of earthquake preparation was poor, particularly with respect to housing. The general management of the hazard was better in Central Italy; for example, less damage to the transport network meant that the emergency services were able to reach the affected areas more quickly.

Preparation is largely about finding ways to reduce the possible impacts of earthquakes and tsunamis. This is sometimes referred to as adjustment, or mitigation. Possible actions fall into:

■ **Construction** – often it is not the earthquake that kills people, but the houses in which they happen to be when the earthquake strikes. With earthquakes, the most dangerous building construction consists of bricks and unreinforced concrete blocks. This is particularly the case where bricks or blocks, held together by mortar, are stacked on top of each other and a roof is laid across the top. Earthquake tremors causes these walls to collapse, and the roof to cave in. There is certainly much that can be done nowadays to make homes safer and more earthquake resistant, such as using lighter materials (timber, aluminium and carbon fibre) and building them on flexible frames with reinforced steel corner pillars.

An inventive solution has recently been put forward by a Japanese company: the levitating or floating home. During stable times, the house sits on a deflated air bag. When sensors detect a tremor, they switch on a compressor within seconds. The compressor pumps air into the airbag, which is also inflated within seconds. The whole house lifts around three centimetres off an earthquake-proof concrete foundation and hovers there for the duration of the earthquake. When the tremors stop, the airbag deflates and the house gently settles back down.

The problem with such solutions is that the cost of 'quake-proofing' homes is high, and well beyond the means of most people in developing countries. For warmer countries, a return to traditional building materials (such as timber, palm and thatch) may be a more cost-effective option.

▶ Figure 3.31: Making high-rise buildings more resistant to earthquakes

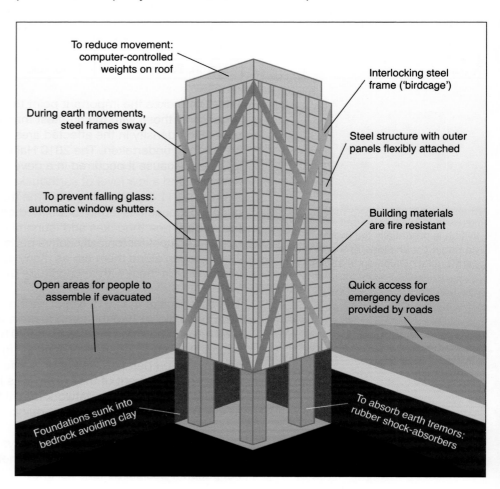

To reduce movement: computer-controlled weights on roof

Interlocking steel frame ('birdcage')

During earth movements, steel frames sway

Steel structure with outer panels flexibly attached

To prevent falling glass: automatic window shutters

Building materials are fire resistant

Open areas for people to assemble if evacuated

Quick access for emergency devices provided by roads

Foundations sunk into bedrock avoiding clay

To absorb earth tremors: rubber shock-absorbers

SKILLS ▶ REASONING

ACTIVITY

Discuss, in groups, why it might be better to use terms such as 'increasing earthquake resistance' rather than 'quake-proofing'.

As buildings become bigger and taller, other construction techniques have been developed to make them more resistant to earthquakes. Figure 3.31 shows some of the building techniques now widely used. However, as with homes, these solutions are expensive and require access to modern building machinery and materials.

In an emergency situation, access to affected areas is a priority. The trouble is that roads, railways and airports are often also seriously damaged. There is not a lot that can be done by way of preparation, save 'quake-proofing' bridges and tunnels and installing systems that bring trains to an immediate halt. Basic services (such as gas, electricity, water and sewage) can be made safer by using flexible piping where possible. Automatic shut-off systems should also be used to reduce the risk of leakages.

- **Warning systems** – as previously mentioned, it is impossible to predict precisely the occurrence of earthquakes. However, once an earthquake has started, equipment such as seismometers, accelerometers, computers and satellites are able to raise the alarm and transmit a warning to those areas that are likely to be affected. Such alarms are most useful to the emergency services. Linked to these warning systems are plans for the organised evacuation of people from affected areas.

  By comparison, warning systems for tsunamis are proving more effective, largely because the waves of seawater move much more slowly than earthquake tremors. Figure 2.39 (on page 60) shows that we can make use of communication and GPS satellites to track their movement and to transmit warnings to those coastal areas that are threatened.

(see Figure 3.15 on page 77)

- **Remote sensing and GIS** – satellite images are increasingly being used in connection with earthquakes. For example, images of the affected area immediately after the earthquake can provide valuable information for search and rescue operations. In earthquake zones, satellite images are also part of the remote sensing that puts together geographic information systems (GIS) about various aspects of the earthquake hazard. For example:

  - mapping the degree of seismic (earthquake) risk (see Figure 3.15 on page 77)
  - detailing the locations of settlements, transport networks and economic activities within those high-risk areas
  - identifying areas where landslides are likely to be triggered by an earthquake

This sort of information is useful as part of earthquake preparation. It can also prove invaluable during the emergency phase (see Section 3.9).

- **Education** – the aim here is to make sure that people know what they should do both during and after an earthquake (Figure 3.32). This includes knowing where to find safe open spaces and locations beyond the reach of earthquakes. In Japan, and in many other high-risk countries, earthquake drills are commonplace in schools and colleges.

ACTIVITY

Discuss, in groups, the advantage of producing GIS in the form of detailed maps of settlements.

CHECK YOUR UNDERSTANDING

Check that you understand the difference between hazard prediction and preparation.

▲ Figure 3.32: An earthquake advice pamphlet

3.9 RESPONDING TO HAZARDS

Having prepared for an earthquake, let us now look at what happens once the hazard strikes (Figure 3.33).

EMERGENCY RESPONSE

Immediately after the event, there is an emergency response. Later, there is a review response in which the whole natural hazard event is examined and questions are asked. What happened? What needs to be done to restore the disaster area? What needs to be done so that next time the damage is less and the death toll lower?

▶ **Figure 3.33: Sequence of events following a hazard event**

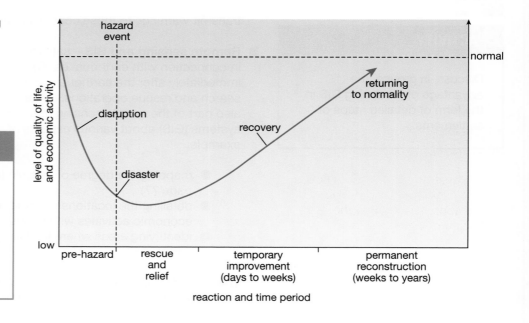

SKILLS INTERPRETATION

ACTIVITY

In your own words, briefly define these terms that appear in Figure 3.33:

■ disruption
■ disaster
■ recovery.

▲ **Figure 3.34: Rescuing survivors from a collapsed building**

Figure 3.33 shows the sequence of events immediately following a hazard. First of all, emergency services must be able to identify quickly the worst hit areas and access them. Time is critical if injured and trapped people are to be rescued. Different types of emergency services will be involved. There will be teams specialising in:

■ releasing people and bodies trapped in collapsed buildings (Figure 3.34)
■ using lifting gear and diggers to clear away rubble and storm surge debris
■ restoring basic services such as water, sewage, gas, electricity and communications
■ providing medical help and counselling victims
■ organising the distribution of emergency rations of food, water and clothing
■ setting up temporary shelters for people made homeless
■ providing transport for emergency supplies – this is often done by the armed forces.

SKILLS ▶ CRITICAL THINKING

ACTIVITY

In class, discuss how and why each of the following would hinder emergency aid:

- airport and seaport damage
- weak government
- widespread poverty
- poor coordination.

SKILLS ▶ REASONING

ACTIVITY

Suggest **three** other questions that might be asked during the appraisal of an earthquake hazard event.

SKILLS ▶ CRITICAL THINKING

ACTIVITY

If there was no optimism, do you think people would still live in hazardous environments? Give your reasons.

All these services need to be well coordinated. This is often where **emergency aid** breaks down. If the disaster is a major one, it will attract relief from international organisations, such as UN agencies, and voluntary organisations such as Oxfam and the Red Cross. But a lack of coordination will slow the relief effort down. In the case of the 2015 Nepal earthquake (see page 76), an additional factor was the severe damage inflicted on roads leading to the worst affected areas.

After the emergency has been dealt with, the next stage of recovery (Figure 3.33) involves deciding what needs to be done to restore the disaster area back to normal – if it is decided that this is the right thing to do. The area may be believed to be too high-risk so that it should be abandoned. This is where organisations such as the World Bank can help in the recovery phase by providing loans to rebuild homes, businesses and **infrastructure**.

APPRAISAL AND LONG-TERM PLANNING

The final stage is appraisal (Figure 3.33). This is the 'inquest', looking back at the disaster and assessing how well or otherwise the emergency operations worked. The appraisal considers whether anything more could be done to reduce the impact if a similar event were to occur again (long-term planning). Attention will focus on the answers to questions, such as:

- Do the risks need to be reassessed?
- Do the warning systems need to be improved?
- Are the evacuation procedures adequate and is everyone aware of what they should do during the emergency phase?
- Can the arrangements for the delivery of emergency aid be improved?
- Can buildings in the area be made more hazard-resistant?
- Are our maps of hazard risk up to date, and is the public aware of them?
- Should particular areas be given greater protection, perhaps in the form of strengthened sea walls?
- Should settlements be relocated away from high-risk locations?
- Is the rebuilding programme fit for purpose – are the buildings sufficiently hazard-proof and located in areas of low risk?

The answers to these questions should guide any long-term planning aimed at reducing hazard risk and its damage potential. It is also at this stage that the United Nations International Strategy for Disaster Reduction (UNISDR) can offer technical advice.

People may be rather optimistic when it comes to natural hazards. First, there is often a feeling that 'lightning will never strike here'. Secondly, there is the belief that 'lightning never strikes the same place twice'. Without that optimism, people would find it difficult to live with the hard truth that natural hazards can:

- strike at any time
- occur in virtually any place, as well as many times in the same place
- be more devastating than previous hazards.

Surprise is one of the nastier aspects of natural hazards, particularly earthquakes.

This section ends with a case study of earthquakes in an emerging country: India. Unlike the earlier case studies of Nepal (page 76) and Italy (page 77), the focus is not on how specific earthquake events were managed. Instead, the focus is on a particular type of earthquake preparation that is as vital as installing warning systems or making sure that new constructions are earthquake resistant.

CASE STUDY: PREPARING FOR EARTHQUAKES IN INDIA

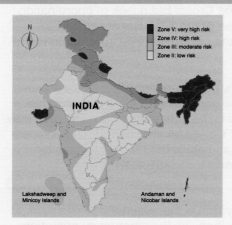

▲ Figure 3.35: Earthquake risk zones of India

Zone V: very high risk
Zone IV: high risk
Zone III: moderate risk
Zone II: low risk

SKILLS INTERPRETATION

ACTIVITY

Look at Figure 3.7 (on page 70). Which two tectonic plates are responsible for the very high risk of earthquakes in the Andaman and Nicobar Islands?

Preparing for earthquakes is a huge challenge for India, because:

■ it is a country where half the land area is prone to damaging earthquakes (Figure 3.35). A number of its largest cities are located in Zones 4 and 5 The north-eastern region of India has a record of earthquakes with a magnitude of 8.0 or more. The main cause of these earthquakes is the movement of the Indo-Australian plate towards the Eurasian plate, at a rate of about 50 mm per year (see Figure 3.7 on page 70)

■ it is a vast country, with a population of around 1.3 billion living at an average density of 400 persons per km^2

The 2016 Nepal earthquake was a wake-up call for its larger and more developed neighbour India. A National Disaster Management Authority has been set up, along with a National Disaster Response Force. In 2014, India began to address a challenge that is widespread in all earthquake zones, and certainly in Nepal and Italy. In India, the challenge is known as 'seismic retrofitting'. This involves modifying existing buildings so that they meet present-day standards of earthquake resistance. Hospitals and schools are given a priority, but blocks of flats should also be included. However, this still leaves a huge backlog of millions of unsafe homes. Retrofitting is also costly, though loans are available. A major problem with the Indian scheme in particular is that there is little official checking that buildings have been brought up to the required standard.

▶ Figure 3.36: Global earthquakes 2000-2016

SKILLS INTERPRETATION (GRAPHICAL)

ACTIVITY

In groups, identify what you think are the four most important 'messages' of Figure 3.36. In a plenary session, compare the group conclusions.

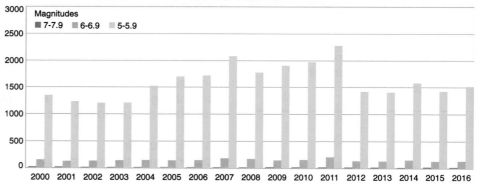

EARTHQUAKES 2000-2016

Magnitudes
■ 7-7.9 ■ 6-6.9 ■ 5-5.9

In summary: even with the best preparation and emergency procedures, earthquakes will continue to cause much destruction and many fatalities. Although the number of earthquakes that occur across the world varies from year to year (Figure 3.36), the long-term trend shows no change in earthquake frequency. Earthquakes remain the most hazardous and destructive of the three hazards examined in this chapter. They occur in more parts of the world than either volcanoes or tropical cyclones, and they cannot be predicted. The destructive power of high-magnitude earthquakes is immense and unrivalled.

CHAPTER QUESTIONS

END OF CHAPTER CHECKOUT

INTEGRATED SKILLS
During your work on the content of this chapter, you are expected to have practised in class the following:

- using world maps to show the distributions of different hazards
- using weather and storm surge data to calculate Saffir-Simpson magnitude
- using social media sources, satellite images and socio-economic ` datato assess hazard impacts
- using maps to the track the movement of tropical cyclones
- using hazard maps.

SHORT RESPONSE

1 Identify **one** potential hazard associated with volcanic eruptions. [1]

2 Identify **two** social impacts of a tropical cyclone. [2]

3 Give **two** examples of emergency aid. [2]

4 Name the hazard measured by the Saffir-Simpson Scale. [1]

LONGER RESPONSE

8th **5** Explain how tsunamis are formed. [4]

7th **6** Compare the primary and secondary impacts of an earthquake. [8]

5th **7** Explain what is meant by the term 'risk assessment'. [4]

8th **8** Analyse how scientists are becoming better at predicting tropical cyclones. [8]

EXAM-STYLE PRACTICE

1 Identify **one** type of climatic hazard. [1]
A A lava flow
B A landslide
C A tropical cyclone
D A tsunami

2 a) State **one** way of measuring an earthquake. [1]

b) Name **one** way people can prepare for an earthquake. [1]

c) Study Figure 3.7 (on page 70). Identify the type of plate margin associated with the Himalayas. [2]

8th **3** Suggest **two** reasons why volcanic eruptions are often less hazardous than earthquakes. [4]

4 Explain the formation of a tropical cyclone. [3]

5 Study Figure 3.23 (on page 83). Identify the country which experienced the third greatest losses. [1]

7th **6** Explain why IGOs are important when it comes to recovering from a hazard event. [4]

8th **7** Analyse why people continue to live in high-risk environments. [8]

[Exam-style practice total 25 marks]

4 ECONOMIC ACTIVITY AND ENERGY

LEARNING OBJECTIVES

By the end of this chapter, you should know:

- The activities that are typical of each of the four economic sectors, and the reasons why employment structures change with development

- The factors affecting the location of economic activity and how these factors change over time

- The reasons for the change in the numbers of workers employed in each of the economic sectors

- The impacts of shifts in the economic sector of specific countries at different stages of development

- The causes and characteristics of informal employment

- Different theories about the relationship between population and resources

- The reasons for the rise in the demand for energy

- The relative merits of different sources of energy

- How energy can be used in a sustainable way

This chapter is about a range of economic activities that provide people with work. They also help countries to develop and become more prosperous. A feature of many economic activities today is that they are changing their locations, both within and between countries. Economic activities and lifestyles in the modern world are demanding more and more energy. The question is how to generate this energy – by using non-renewable or renewable resources?

4.1 ECONOMIC SECTORS AND EMPLOYMENT

The production of food (farming), the making of goods from raw materials (manufacturing), the provision of services, and of information and expertise are essential to our lives. The term 'economic activity' is used to describe any type of undertaking within this broad range. Each and every economic activity does at least three things:

- it creates jobs (employment)
- it generates income (wealth)
- it produces something for sale or **consumption**.

Most economic activities are also driven by the need for work and to make a living.

Economic activities can be grouped according to what they produce and the types of jobs they offer. Each group is known as an economic sector (Figure 4.1). Three are widely recognised throughout the world: primary, secondary and tertiary. A relatively new, fourth sector – quaternary – has been added in most developed and some emerging countries.

THE FOUR ECONOMIC SECTORS

Let us look at each of these sectors and their typical economic activities.

- **Primary sector** – working with natural resources. The main activities are farming, forestry, fishing, mining and quarrying.
- **Secondary sector** – processing things such as food or minerals (for example, iron ore), making things by manufacturing (for example, microchips), assembling (for example, cars) or building (for example, houses).
- **Tertiary sector** – providing services. These include services that are commercial (for example, retailing and banking), professional (for example, solicitors and accountants), social (for example, schools and doctors), entertainment (for example, restaurants and cinemas) and personal (for example, hairdressers and fitness trainers). Public and private transport is also included in this sector.
- **Quaternary sector** – concerned with information and communications (ICT) and research and development (R & D). Universities are an important part of this sector.

The relative importance of the sectors in a country's economy is a good indicator of the level of economic development. Generally, the economy of a developing country relies heavily on the primary sector, while the economy of a developed country depends most on the tertiary sector. How do we measure the relative strength of the sectors? We need to use the same measure so that countries may be reliably compared. There are two different measures.

The first measure is employment. The sectors are compared in terms of the percentage of the total workforce that they employ. A pie chart is a good way of showing this information (see Figure 4.13 on page 110). The second measure is based on how much each sector contributes to the overall economic output of a country – their percentage of either **gross domestic product** (GDP) or **gross national income** (GNI). Again, a pie chart is a useful way of showing this measure.

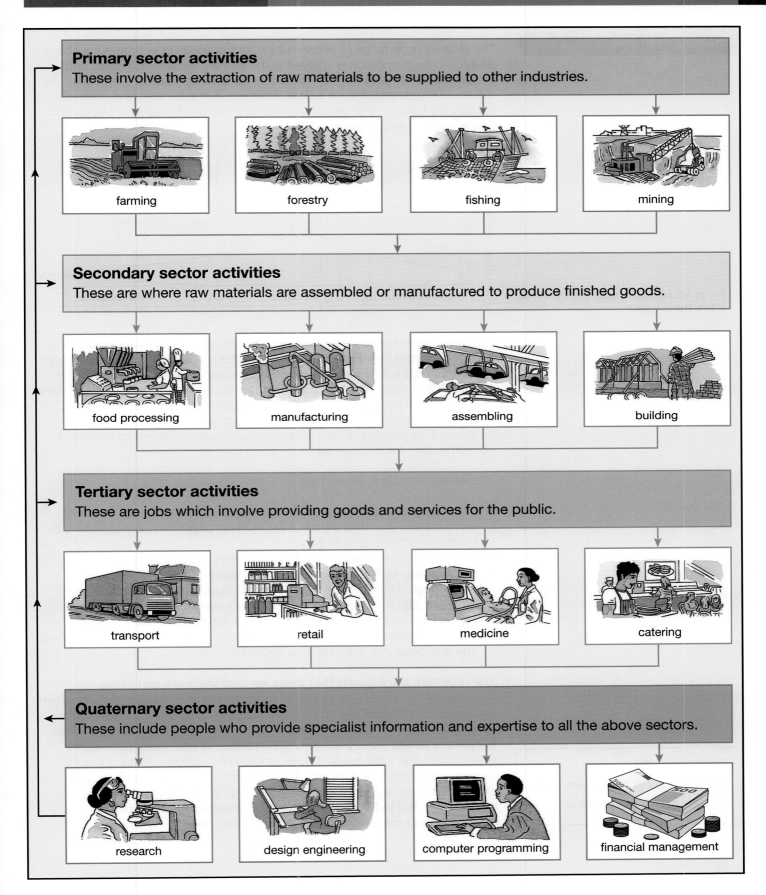

▲ Figure 4.1: Economic sectors and their activities

CHANGES OVER TIME

The relative importance of economic sectors changes over time. As a country develops, the proportion employed in the primary sector decreases, and the proportion employed in the secondary and tertiary sectors increases. As the economy develops even further, numbers in the primary and secondary sectors fall further. After the tertiary sector becomes the largest employer, a quaternary sector begins to emerge.

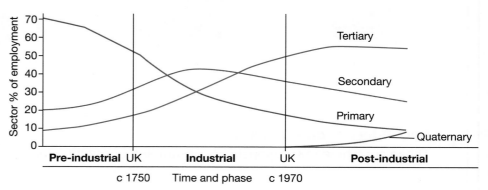

▶ Figure 4.2: Economic sector shifts in the UK (Clark-Fisher model)

Figure 4.2 shows how the sectors change with time over three phases. This is known as the Clark-Fisher model. The dates of the UK's progress through these phases are also shown. The critical phase is the industrial one. This occurs when manufacturing (the secondary sector) becomes more important than the primary sector, in terms of both employment and contribution to GDP. Cottage industries are replaced by mechanised industries housed in factories. This process of **industrialisation** is accompanied by important economic and social changes.

The three phases are shown here.

■ Pre-industrial phase – the primary sector leads the economy and may employ more than two-thirds of the working population. Agriculture is by far the most important activity.
■ Industrial phase – the secondary and tertiary sectors increase in productivity. As they do so, the primary sector declines in relative importance. The secondary sector peaks during this phase, but rarely provides jobs for more than half of the workforce.
■ Post-industrial phase – the tertiary sector is now the most important sector. The primary and secondary sectors continue their relative decline. The quaternary sector begins to appear.

The important point here is that these sectoral shifts are part of the development process. The development pathway starts as agriculture, which becomes mechanised, becomes more commercial and shifts away from **subsistence farming**. This releases labour to take on other forms of work. People are free to move from the countryside and into urban settlements where traditional craft industries are replaced by factories making goods for expanding markets. The wages earned in the factories lead to more disposable income and much of this is spent on a widening range of services. So, the tertiary sector gradually becomes dominant both in terms of employment and generating economic wealth.

DID YOU KNOW?

Look at Part 9.3 for more information about the development pathway.

▶ **Figure 4.3: Research: a major activity in the quaternary sector**

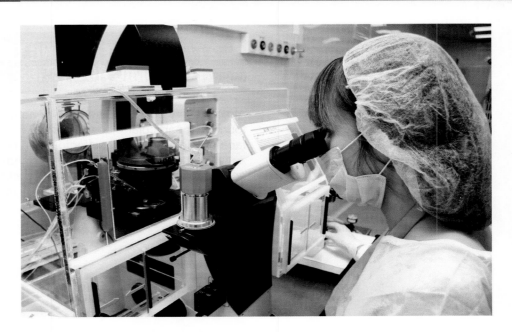

The quaternary sector appears when leading countries, and major cities within those countries, find that to keep 'ahead of the pack', they need to invest in higher education, research and development and new technologies (Figure 4.3).

▼ **Figure 4.4: The global distribution of economic development**

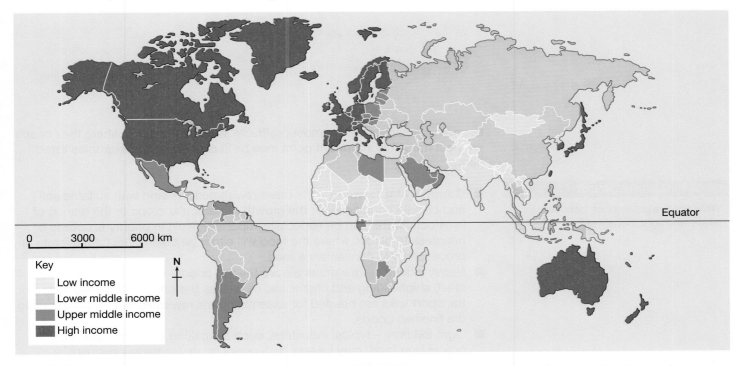

Equator

```
0        3000      6000 km
```

Key
- Low income
- Lower middle income
- Upper middle income
- High income

N

SKILLS ▶ PROBLEM SOLVING

ACTIVITY

Look at Figure 4.4. Into which category do known emerging countries such as Brazil, China, India and Russia fall?

The balance of economic sectors also changes from country to country, depending on the level of development. Figure 4.4 shows the global distribution of four widely-recognised levels. Developing countries fall in the low-income category and developed countries in the high-income category. In between those two categories are the emerging countries. All this emphasises the important point that development and wealth are closely related.

4.2 FACTORS AFFECTING THE LOCATION OF ECONOMIC ACTIVITIES

Almost every economic activity – whether it is a mine, a factory or a shop – is found in a particular place for good reasons. Those reasons are called location factors (Figure 4.5). These factors can be described as the needs of the activity. They are space (most often land), raw materials, labour (workers) and access to markets (good transport, customers). The mix of needs (their relative importance) varies from activity to activity.

▶ Figure 4.5: Factors influencing location

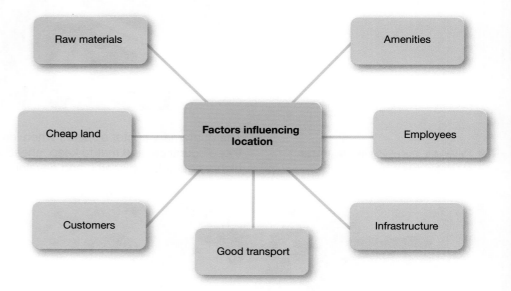

Economic activities will be most profitable in those locations where their needs are best met. This important point may be illustrated by the examples listed here.

CHECK YOUR UNDERSTANDING

What are the differences between heavy and light industries?

- **Commercial farming** – the basic raw material is land with suitable soil and climate conditions for the growth of particular crops or the rearing of livestock. Water can be especially important. Next in priority is access or nearness to places where the food will either be taken in its natural state or processed into convenience food.
- Heavy industry – raw materials are the top priority for industries such as steel, shipbuilding and chemicals, as well as the supply of energy. Good transport links are needed for assembling the raw materials and distributing the finished goods.
- Light industry – typical industries, such as making electrical goods or processing food, may be less dependent on raw materials but usually consume large amounts of energy. Some of the more advanced industries also depend on the availability of skilled labour.
- Retailing – many services within the tertiary sector have to be readily accessible to customers. After all, if there are no customers, there would be few services. But since services are typically labour-intensive, they also need to be able to recruit large numbers of workers.
- Research and development – this is an important quaternary sector activity. It relies most on a highly-skilled workforce and access to the latest research being undertaken, usually in universities.

SKILLS ▶ SELF DIRECTION

ACTIVITY

Look at the CBD of a town or city near you. Make a list of the main types of service available there.

SKILLS ▶ REASONING

ACTIVITY

Produce a diagram similar to Figure 4.6, but which shows the downside of these developments in the urban fringe. You might look at Chapter 6.

Accessibility is a key location factor with most economic activities – access to raw materials, markets or labour. This explains why so many services are concentrated in the central areas of towns and cities. These concentrations are known as central business districts (CBDs) and are found in urban areas worldwide.

The CBDs are accessible because this is where the urban transport networks, both public and private, converge. Transport networks serving surrounding areas and linking to other urban settlements also converge here. Remember that the customers of services in the CBD are not just urban residents. They come from all those places lying within the city's (or town's) sphere of influence.

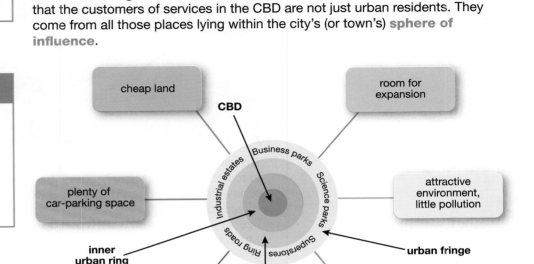

▶ Figure 4.6: The locational attractions of the urban fringe

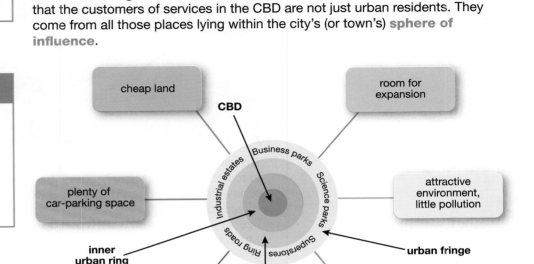

▲ Figure 4.7: An urban fringe retail park, Zaragoza, Spain

A feature of cities in the developed world over the last 25 years has been decentralisation. People and businesses, particularly tertiary activities, have been moving out from the CBD and inner city to the suburban ring and just beyond into the urban fringe. Figure 4.6 shows some of the locational attractions of the urban fringe for both tertiary and quaternary activities. The signs of this happening around the urban fringe include the appearance of certain buildings.

- ■ Superstores and retail parks – large areas with adjacent car parks occupied by either one huge hypermarket or a number of retailing companies in separate buildings (Figure 4.7); these developments often serve customers drawn from more than one town or city.
- ■ Industrial estates – areas of modern light and service industries with a planned layout and purpose-built road network.
- ■ Business parks – areas created by property developers to attract firms needing office and retail accommodation rather than industrial units; they often include leisure activities such as bowling alleys, ice rinks and cinemas.
- ■ Science parks – usually located close to a university or res   earch centre with the aim of encouraging and developing high-tech industries and quaternary activities (Figure 4.8 and see also Part 6.7 on page 173).

▶ Figure 4.8: A modern science park, Cambridge, UK

CHECK YOUR UNDERSTANDING

Give some examples of high-tech industries.

DID YOU KNOW?

For more information about tourism, see Parts 8.6 and 8.9.

Not all tertiary activities are confined to urban areas. One obvious exception is tourism. This is probably one of the most important tertiary sector activities today. It employs millions of people worldwide and creates huge amounts of economic wealth. The coast is a powerful location factor, because of its scenery, fresh air and water sport opportunities, from swimming to sailing. Much tourism is also located in rural areas in, for example, national parks and other protected areas.

4.3 CHANGES IN SECTOR EMPLOYMENT

The Clark-Fisher model (Figure 4.2 on page 100) is based on the idea that, as a country develops, the relative importance of the economic sectors changes in terms of:

- their contribution to a country's economic effort (GDP or GNI)
- the percentage of a country's labour force employed in each sector.

These shifts are the result of a number of factors and processes. All these are related in some way to the broad development process (see Chapter 9).

SKILLS REASONING

ACTIVITY

How does the relative importance of the economic sectors change with development?

CAUSES OF CHANGE

Location factors change. This often results in economic activities also changing their locations. Any changes will affect employment. The relocation of a factory could mean a loss of secondary sector jobs in the original location and a gain in the new one. But there are other reasons why the numbers of people employed in each economic sector change over time. We will look at five of these reasons.

RAW MATERIALS

The sources of raw material, for example minerals such as iron and copper, often become exhausted. When this happens, manufacturers may change their location. Most manufacturing is no longer tied to the coalfields, as was the case in the UK in the 19th and early 20th centuries. Much of the energy needed by manufacturing now comes from electricity. Thanks to grid networks, electricity can be supplied to almost any location. The same is true of oil and gas distributed by pipelines and tankers.

▶ Figure 4.9: Mechanisation in India of:
a) agriculture, and b) manufacturing

ACTIVITY

What have been the impacts of mechanisation on agriculture?

a

b

NEW TECHNOLOGY

Advances are constantly made in technology. Many of these advances impact directly on the economic sectors. For example, the mechanisation of agriculture and manufacturing has, at different times, reduced their demand for labour (Figure 4.9a and b). The labour no longer required by farming moved into manufacturing. Later, the labour made redundant by industrial machinery has found alternative work in the tertiary sector.

Technology advances have also had a major impact on transport, greatly reducing the **friction of distance** (Figure 4.10). It is now possible to move people and goods much more quickly and relatively cheaply. Thanks to modern **communications**, the transfer of information around the world is almost instantaneous. The net effect is that places have become closer and better connected.

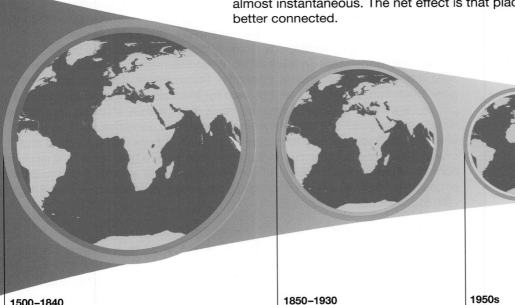

1500–1840
Best average speed of horse-drawn coaches and sailing ships was 10 mph

1850–1930
Steam locomotives averaged 65 mph and steam ships 36 mph

1950s
Propeller aircraft 300-400 mph

1960s
Jet passenger aircraft 500-700 mph

▲ Figure 4.10: Advances in transport reduce the friction of distance

But technological advances are not only about introducing new ways of doing things. Its most advanced forms have created new industries, called high-tech industries, such as aerospace, biotechnology, robotics and telecommunications.

Technological advances have also created new products and services such as the smart phone, tablet, laptop and MP3 player. These have all appeared within the last 20 years. Think, too, of all the new services connected with ICT – broadband service providers, website designers, mobile phone networks, software programmers and the servicing of PCs and laptops.

▶ **Figure 4.11: Factors encouraging the global economy**

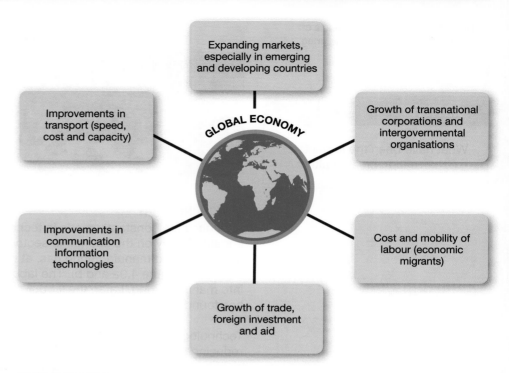

GLOBALISATION

Globalisation describes the process by which the countries of the world are being gradually drawn together into a single **global economy** by a growing network of links and organisations (Figure 4.11). The advances in transport and communications technology have contributed to countries becoming increasingly dependent on each other. What this means is that all places can concentrate on their economic strengths. For example, Kenya is now able to grow fresh fruit, vegetables and flowers for markets in Europe. Competitively-priced cars made in South Korea are easily and cheaply shipped around the world. International call centres are now located in countries such as India, where English is a second language and labour is both well-educated and cheap. Countries such as Sri Lanka and Cyprus have been able to share in the boom in international tourism. All such developments affect the distribution of employment from place to place, and from sector to sector. See Part 8.1 (page 204) for more about globalisation.

GOVERNMENT POLICIES

Any responsible government will be concerned about their country's economy and its future prospects. The degree of government intervention in the economy varies from minimal in capitalist countries to complete control in communist countries. In the UK, policies have aimed at supporting agriculture and encouraging services to compensate for the loss of jobs in manufacturing. In China, there has been a concentrated effort to expand the secondary sector and to produce manufactured goods for a global market.

DEMOGRAPHIC AND SOCIAL CHANGE

This is about people. Clearly, populations change over time. Mostly they grow. This raises the demand for a range of goods and services such as food, manufactured home appliances and schools and medical services. The growth of these goods and services, in turn, will boost the economic sectors, but to varying degrees. Population growth also means more workers and this can encourage the growth of economic activities needing a plentiful supply of labour.

▶ Figure 4.12: The cycle of growth in the tertiary sector

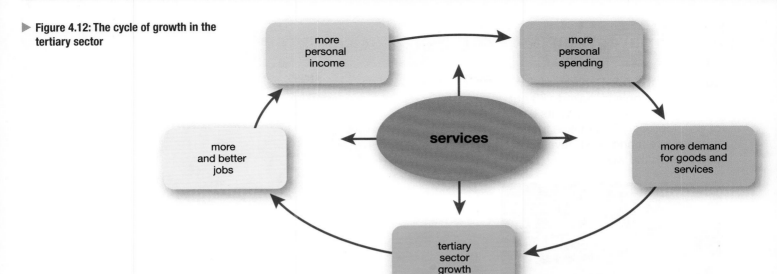

Other demographic changes occur as a country moves along the development pathway. This is illustrated by what happens in the tertiary sector:

■ people earn more money and have money to spend in the shops on basic requirements such as food and clothing
■ after they have bought the basics, people have more money left (**disposable income**) to spend on non-essential goods and services such as entertainment, holidays, eating out and recreation
■ people's preferences change and this impacts on the tertiary sector; for example, people in HICs now prefer to shop on a weekly basis at superstores and retail parks rather than making daily visits to a local shop.

All these changes will clearly encourage a cycle of growth in the tertiary sector by creating more services and more jobs (Figure 4.12).

SKILLS PROBLEM SOLVING

ACTIVITY

What is significant about disposable income?

4.4 SECTOR SHIFTS IN THREE COUNTRIES

Sector shifts are changes in the relative importance of the economic sectors that take place as countries develop. The sequence has been described on page 104. See also Figure 4.2 on page 100.

The case studies that follow will look at the impacts of sector shifts in three countries at different stages of development:

■ Ethiopia – a developing country in the pre-industrial phase
■ China – an emerging country in the industrial phase
■ the UK – a developed country now in the post-industrial phase.

Together they represent different points along the development pathway.

CASE STUDY: ETHIOPIA – A PRE-INDUSTRIAL COUNTRY

Located in the sub-Saharan region of Africa, Ethiopia is ranked 173 out of 186 countries in terms of its level of development. This makes it one of the poorest and least developed countries in the world.

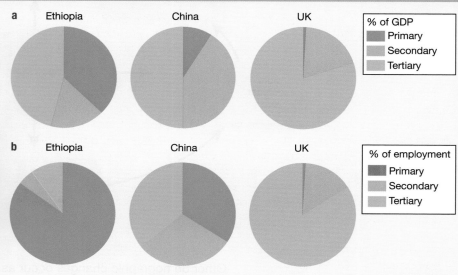

▲ Figure 4.13: A farming landscape in Ethiopia   ▲ Figure 4.14: A comparison of three countries: a) by GDP, and b) by employment (2016)

Figure 4.14 shows that Ethiopia's economy is focused on the primary sector. The primary sector accounts for 75 per cent of all employment but slightly less than 50 per cent of GDP. This difference in percentage values reflects the fact that agriculture in Ethiopia is mainly subsistence farming (Figure 4.13). But this is beginning to change. Coffee has for a long time been the main commercial crop. The cultivation and export of oil seeds and flowers are now on the rise. An obvious negative impact is that this growth of commercial agriculture is taking over land that was once used for subsistence farming. This, in turn, is increasing food shortages in a country that has regularly suffered from **famines**. It also influences people to move to Addis Ababa, the capital city, and to other urban centres, in search of a way to make a living. Another change within the primary sector is the growth of gold mining. This is creating jobs and earning foreign currency.

Up until now, the secondary sector (manufacturing) has played a small part in Ethiopia's economy. But that is beginning to change. The processing of agricultural products has now been joined by the making of textiles and leather goods as the country's leading industries. Exports are growing. The government is investing in the country's infrastructure, particularly in improving the road network and the supply of energy. Foreign investors are beginning to take an interest in Ethiopia. The availability of affordable labour might prove a strong attraction and persuade Transnational Corporations (TNCs) to set up factories there.

Over 75 per cent of Ethiopians live in rural areas, but this percentage is beginning to fall. Rural–urban

ACTIVITY

Besides affordable labour, what else might attract investors to set up businesses in Ethiopia?

migration is increasing. Cities and towns are few and far between. This helps to explain why there is so little employment in the tertiary sector. A poor, largely rural population has little or no money to spend on urban services. However, since services are relatively more valuable than commercial crops, the small tertiary sector accounts for almost as much of the country's GDP as the primary sector (Figure 4.14). Again, change is happening in the tertiary sector. The African Union and the UN Economic Commission for Africa have set up their headquarters in Addis Ababa.

So, Ethiopia is beginning to make economic progress. Its economy is still rooted in the primary sector. There are shifts taking place, but perhaps more within the individual sectors than between them. The shifts are improving the **quality of life** and living standards of some Ethiopians. The most obvious negative impact is the displacement of subsistence farmers by the growth of commercial agriculture.

ACTIVITY

Find out possible reasons that are making development difficult for Ethiopia.

CASE STUDY: CHINA – A RAPIDLY EMERGING COUNTRY

China is a giant of a country in terms of its area and its population. Over the last 10 years, its economy has grown very quickly. It is now the largest economy in the world. In per capita terms, however, it is still a lower middle-income country.

A shift from the primary to the secondary sector has driven much of China's recent economic development and prosperity. The shift has been the result of a decision by the Chinese government that the country should become part of the global economy. Previously, it had only been part of the communist world. It traded mainly with what was the Soviet Union and the countries of Eastern Europe. The secondary sector now accounts for half the country's GDP, but only a quarter of the labour force. The industrial success has been based mainly on the availability of cheap labour and energy.

It is interesting to note that agriculture in the primary sector still employs large numbers of workers. However, this sector's contribution to GDP is shrinking fast. But China's growing urban population still needs feeding. There is no doubt that the rural population is being left behind. There is a widening gap between them and urban people in terms of quality of life. The shift to the secondary sector has benefited many people. Higher wages and living standards are two obvious examples. The modern architecture of megacities, such as Shanghai and Beijing, is impressive (Figure 4.15). The streets are crowded with well-dressed young people. Everyone seems to have money to spend on the latest technology.

However, there are also costs. Perhaps the most noticeable is the serious pollution produced by the heavy industry and urban traffic. While the economy has opened up, access to the internet is still carefully controlled.

SKILLS   DECISION MAKING

ACTIVITY

Discuss, in groups, why China might have a food supply problem.

SKILLS   SELF DIRECTION

ACTIVITY

What sort of goods are being produced by Chinese manufacturers? Do you have any goods in your home that were made in China? If so, what?

SKILLS   REASONING

ACTIVITY

Why do you think access to the internet might be restricted in China?

◀ **Figure 4.15: A busy street in Shanghai**

CASE STUDY: THE UK – A POST-INDUSTRIAL COUNTRY

The UK was the world's first industrial nation. It led the Industrial Revolution. Fifty years ago, manufacturing produced 40 per cent of the country's economic wealth and employed one-third of the workforce. Today, it produces slightly less than 25 per cent of the wealth and employs less than 20 per cent of the workforce. As a result of the **global shift** in manufacturing (see Part 8.1 on page 214), the country has experienced **de-industrialisation** (Figure 4.16). Many of the goods once manufactured in the UK are now made in China, India and other MICs and LICs.

The UK's economy today is very much a service-based economy. Figure 4.14 (on page 108) shows that the tertiary sector provides jobs for 80 per cent of UK workers and creates 75 per cent of the national economic wealth. These figures for the tertiary sector also include those for the up-and-coming quaternary sector. Because it is often hard to distinguish between quaternary and tertiary sector work, we can only guess what the quaternary sector's contribution to the economy is – possibly 10–15 per cent.

▶ **Figure 4.16: De-industrialisation of the Potteries**

It is interesting to note the place of agriculture in the UK economy. UK farming produces about 60 per cent of the country's food supply. The low labour percentage reflects the high level of mechanisation; while a low GDP percentage reflects the low price of farm products relative to manufactured goods and services. There is no doubt that the sector shift from the secondary to the tertiary sector has had its social costs. De-industrialisation has meant workers being made redundant, rising unemployment and the challenge of finding new jobs. The search for work has often led to families being uprooted from their homes. It was fortunate that the decline of manufacturing coincided with growth in the tertiary sector.

De-industrialisation also had huge environmental costs. Large urban areas were laid waste as factories closed and were demolished (Figure 4.16). Industrial towns and cities in the north of the UK have had to reinvent and re-image themselves. The survival of these settlements has been both difficult and expensive.

SKILLS DECISION MAKING

ACTIVITY

Discuss, in groups, whether or not you think that the de-industrialisation of the UK has all been bad news.

CHECK YOUR UNDERSTANDING

Identify **three** possible reasons for de-industrialisation.

The pie charts in Figure 4.14 on page 108 will help you understand what is meant by sector shifts. The relative importance of the economic sectors changes as you read across from Ethiopia to the UK.

In the three case studies, it is clear that some of the UK's manufacturing has moved to China, causing sector shifts in both countries. Will some of China's manufacturing eventually move to one of today's LICs, such as Ethiopia? It is possible, particularly if Chinese workers continue to expect higher wages. This will make Chinese labour more expensive. It could well persuade manufacturers to move to locations where labour is cheaper. This would be bad news for China, but possibly good news for Ethiopia.

4.5 INFORMAL EMPLOYMENT

So far, we have recognised four different sectors. In some parts of the world, there is a fifth sector that is not recognised in the official figures produced by governments. It is called the **informal sector**, and is also referred to as the 'black economy' because it is unofficial and unregulated. Yet this sector employs millions of people across the world, especially in developing countries (Figure 4.17).

In some countries, possibly over 70 per cent of the workforce finds work in the informal sector.

▶ Figure 4.17: An estimated global distribution of informal employment

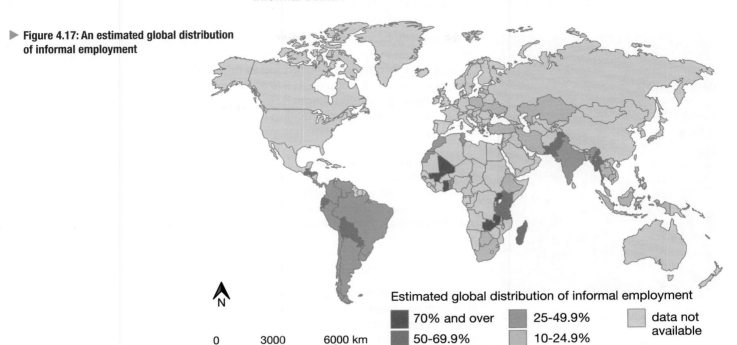

Estimated global distribution of informal employment

- 70% and over
- 50-69.9%
- 25-49.9%
- 10-24.9%
- data not available

0   3000   6000 km

CAUSES

CHECK YOUR UNDERSTANDING

Check that you understand the difference between underemployment and unemployment.

It is important to understand why the informal sector develops. This will help us to understand its characteristics. In many developing countries today large numbers of people are migrating from rural to urban areas. An important driving force behind that migration is the search for work and a regular wage, together with the belief that the quality of life is better in towns and cities. While it is true that there are more job opportunities and higher wages, there are often more people of working age moving into urban areas than there are jobs available. Surplus labour means that there is **underemployment** and unemployment. In this situation, employers can only pay their workers very low wages. Wages are so low that they are not enough for a worker and their family to live on. So, to avoid poverty, many people must find other ways of making a living outside the normal job market.

CHARACTERISTICS

▲ Figure 4.18: Child labour in Dhaka's paratransit

Ways of making a living might involve selling matches or shoelaces on the street, ice-cream vending, shoe-shining, rubbish collecting or scavenging bottles, cans and other types of waste for recycling. Some people are so desperate they resort to begging, petty crime or prostitution. Informal economic activities are mainly within the tertiary sector. Informal employment is closely associated with another common feature of urban areas in developing countries – shanty towns (see Part 6.3).

An interesting group of informal activities falls under the heading of **paratransit** (Figure 4.18). These activities arise because of the inadequate official transport in towns and cities in developing countries. They usually take the form of minibuses, hand-drawn and motorised rickshaws, scooters and pedicabs (tricycles used as taxis). While paratransit activities do well because they meet the demand for cheap urban transport, they frequently add to the problems of **congestion** on already busy, overloaded roads.

CASE STUDY: DHAKA, BANGLADESH

▲ Figure 4.19: Dhaka: – a densely-populated megacity

SKILLS CRITICAL THINKING

ACTIVITY

Make a two-column table and list:

■ the benefits of informal employment
■ the costs of informal employment.

Which column do you think carries more weight?

This megacity has a population of 16 million and is one of the fastest growing cities in the world. Dhaka (Figure 4.10) attracts economic migrants from all over Bangladesh. According to the World Bank, around 25 per cent of the capital's population live in crowded slums. It is by far the most densely populated megacity in the world. We can only guess what percentage of the city's population depends wholly or partly on the informal sector. It could be more than half.

These informal activities certainly have some benefits. For example, they provide a wide range of cheap goods and services that would otherwise be out of the reach of many people. They provide a means of survival. However, because earnings are slow, informal activities do nothing to break the cycle of poverty in urban areas in developing countries.

Other costs associated with the informal sector include:

■ no healthcare or unemployment benefits
■ a high exposure to work-related risks
■ an uncertain legal status
■ discrimination.

Perhaps the worst aspect is the involvement of children in economic activity, rather than formal education. In Dhaka, it is estimated that there are half a million children in the informal sector. Most of them work from dawn to dusk, earning on average the equivalent of 50 cents (US) a day, to help support their families. The jobs range from begging and scavenging to domestic service and working as fare collectors for various forms of paratransit. Dhaka is known as the rickshaw capital of the world.

The informal child workers of Dhaka, as in other cities in developing countries, work in dangerous conditions; they are exposed to hazards such as street crime, violence, drugs, sexual abuse, toxic fumes and carrying excessive loads (Figure 4.18). Their working conditions mean that these children often suffer extremely poor health and a range of development problems. There is little hope that they will break out of the cycle of poverty, particularly if they have no schooling.

4.6 POPULATION AND RESOURCES

The link between this section and the rest of the chapter is that most economic activities involve the consumption of resources and energy. The rate at which resources are consumed is strongly influenced by two processes: population growth and development.

The global population is now past the 7 billion mark and continues to rise rapidly. Global economic growth is about 3 per cent a year. Many of the resources being consumed are not limitless. They are finite and non-renewable. So, with resources running out, what will happen to people and their economic activities? Is the world heading for a catastrophic disaster?

▶ Figure 4.20: Three views on the relationship between population growth and resources

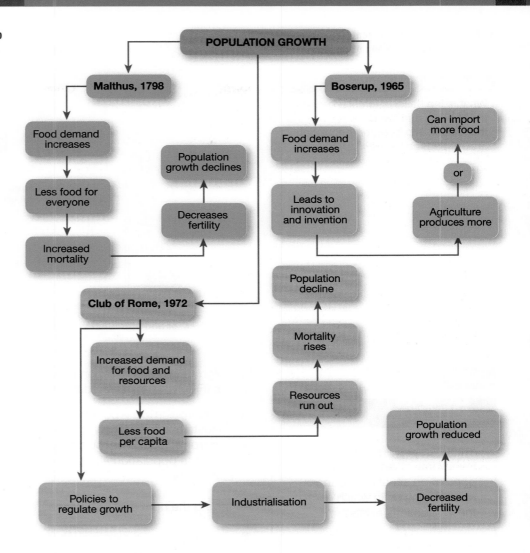

Three different views have been expressed about the relationship between population growth, development and resources (Figure 4.20). With all three, the focus is on food.

- Malthus (1798) was the first person to put forward a very pessimistic view. He argued that population growth proceeded at a faster rate than the increase in food supply. So, there would come a time when there was no longer sufficient food to feed the population. At this point, population growth would stop, either by a lowering of the birth rate (people having fewer children), or a raising of the death rate as a result of famine, disease and war.
- Boserup (1965) based her theory on the argument that increases in population stimulate an improvement in food production. So, developments in technology would solve the problem.
- The Club of Rome (1972) argued that the limits to global population growth would be reached within the next 100 years if population and development continued at the rates of growth seen in the 1970s. However, they suggested that it would still be possible to reduce the trends by means of growth-regulating processes. If this happened, then a sustainable triangular balance between population, development and resources might be reached. Presumably modern technology could also be used to reach the equilibrium.

▶ **Figure 4.21: Three different relationships between population and resources**

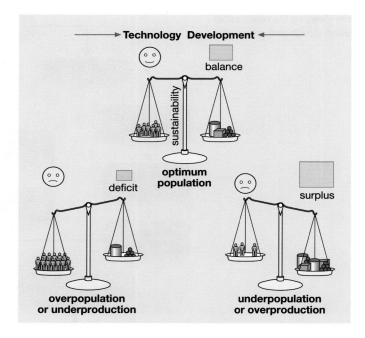

SKILLS PROBLEM SOLVING

ACTIVITY

Identify possible signs or symptoms of each of the three population situations. Can you name countries in which each of these situations exists?

We might think of population and resources as being weighed by a set of scales (Figure 4.21). Three different balance situations are possible:

■ **overpopulation** – population numbers exceed resources; this is an unsustainable situation
■ underpopulation – resources exceed population numbers; a rare situation
■ optimum population – population and resources are in a balance which is sustainable.

Achieving the right sort of balance depends on:

■ controlling population growth, for example by family planning
■ reducing our own resource consumption by using resources much more efficiently and cutting out waste
■ using technology to discover and exploit new resources
■ making sure that development is less based on natural resources.

Perhaps you can see that this list combines some of the ideas of Malthus, Boserup and the Club of Rome as ways of reducing **population pressure** on resources.

4.7 RISING ENERGY DEMAND

In this part of the topic you should recall and make use of what you have learned from your research into different approaches to developing energy resources.

Energy is one of the most important of all the world's resources. We need energy to keep us warm and to cook with. It gives us light and powers industrial machinery and transport. Fortunately, the natural environment provides us with a wide range of energy sources. There is a difference between primary energy and secondary energy. Primary energy describes fuels that provide energy without undergoing any conversion process, for example coal, natural gas and fuelwood. Secondary energy includes electricity, petrol and coke, which are made from the processing of primary fuels. In today's world, electricity is undoubtedly the leading source of energy.

There is another important distinction made in the world of energy. Energy sources such as fossil fuels (coal, oil and gas) are classed as **non-renewable**. Once used up, they cannot be replaced. Newer energy sources such as solar, wind and tidal power are described as **renewable**. They can be used again and again. For this reason, they are sustainable and are likely to play an increasingly important role in the future.

ENERGY DEMAND

The demand for energy across the world is constantly rising (Figure 4.22). This increase in demand is caused by the increase in population and by economic development. The amount of   energy that a country uses is widely used as an indicator (or measure) of its level of development. As a country develops, energy-consuming activities such as manufacturing, provision of services and transport increase in scale and importance. This rising demand for energy will be met by either the country using its own energy resources or importing energy from producer countries (Figure 4.24).

Figure 4.23 shows the global distribution of **energy consumption**. It also shows energy demand. Europe and North America use 70 per cent of the world's energy, although only 20 per cent of the world's population lives there. These areas were the first to experience large-scale economic development. They used their own supplies of fossil fuels to provide the necessary energy for this development. Today, with many of their own reserves falling low or finished, they need to import energy, especially oil, to meet their ever-rising energy demand.

▶ **Figure 4.22: The rising global demand for energy (figures for 2040 are estimated)**

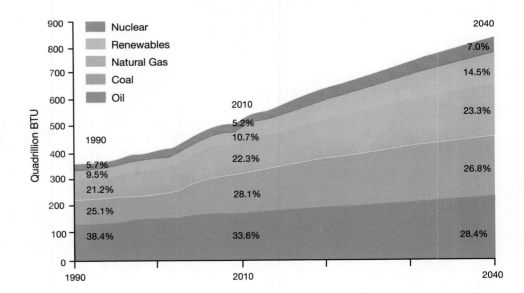

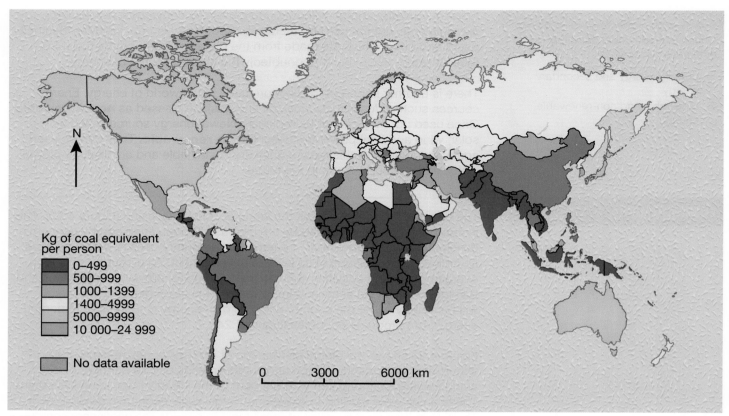

▲ Figure 4.23: Global distribution of energy consumption

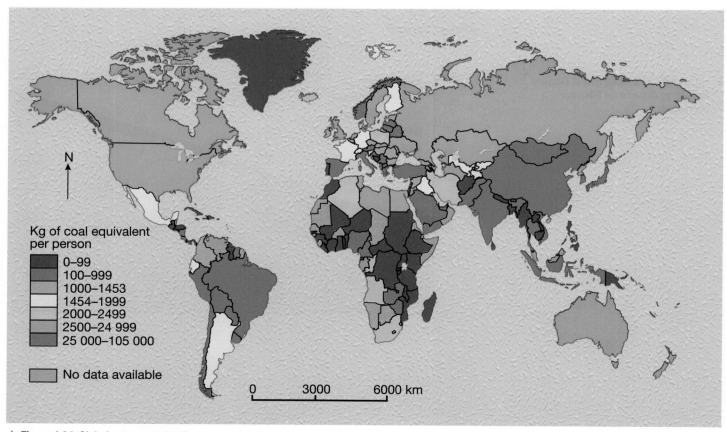

▲ Figure 4.24: Global energy production

A noticeable feature of Figure 4.23 is the relatively low level of energy demand over much of South America, Africa and South and South-East Asia. This mainly reflects a relatively low level of economic development. However, remember that these regions enjoy warm climates and therefore little energy is needed for heating. Cooling is another matter and many HICs use much energy for cooling.

ENERGY PRODUCTION

SKILLS   DECISION MAKING

ACTIVITY

What are the two 'energy worlds'? Which of them do you think is better? Give your reasons.

Three-quarters of the world's energy production comes from three sources: oil, natural gas and coal. All these are non-renewable. The major producers of energy are the USA, Canada, Western Europe, Russia, parts of the Middle East, Australia and New Zealand (Figure 4.24). These areas have reserves of one or more of these main energy sources. By comparing Figures 4.23 and 4.24, we can see that the world's major consumers of energy are also the major producers. Put another way, levels of energy production are low in those countries with low levels of energy consumption and demand – the developing countries. There are two different 'energy worlds'.

ENERGY SECURITY

Energy security exists when a country is able to meet all of its energy needs reliably, preferably from within its own borders (Figure 4.25). The number of such countries is small. Today, most countries face an **energy gap** between energy demand and energy supply. In these countries, the rising demand for energy can only be met by importing energy.

This is well illustrated by the case of the UK. There, and elsewhere, the energy gap is being widened by the deliberate phasing out of fossil fuels as a way of reducing carbon emissions. Unfortunately, renewable sources of energy cannot replace in full all the energy lost if fossil fuels were no longer used. So, the burning of fossil fuels continues. Ironically, the UK still has plenty of coal deposits, but it is now cheaper to import foreign coal. Similarly, there is still oil and gas in the North Sea; but because of its price, more oil and gas is being imported. As a result, the energy gap is widening and the UK is becoming less energy secure.

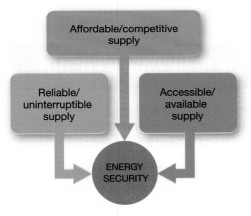

▲ Figure 4.25: Energy security

It is doubtful whether the world as a whole has yet reached the point of having an energy gap or crisis. But individual countries have (Figure 4.26). There are certainly plenty of reserves of coal, oil and natural gas. However, these sources are non-renewable and their widespread use is being questioned. There is a critical need to reduce carbon emissions due to their possible link with global warming. The world's immediate energy problem today is the growing mismatch between the distribution of energy consumption and production. This mismatch is creating not only national energy gaps, but also countries with energy surpluses. These surpluses give those countries huge geopolitical and economic power. They have the ability to hold other countries to ransom. Many would question how that financial power is being used by some of those countries.

SKILLS   PROBLEM SOLVING

ACTIVITY

Study Figure 4.26 on the next page. Explain the link between energy security, energy surplus and energy deficit.

Energy Surplus

- Above 200
- 200-100
- 100-0
- -0-100
- -100-200
- Above -200

Energy Deficit

- No data

△ Major coalfield
◉ Major gasfield
● Major oilfield

0   3000   6000 km

Source: www.mapsofworld.com

▲ **Figure 4.26: The global distribution of energy security**

4.8 RENEWABLE VERSUS NON-RENEWABLE ENERGY

In this part of the topic you should recall and make use of what you have learned from your research into different approaches to developing energy resources.

NON-RENEWABLE SOURCES

The world's energy resources can be divided into non-renewable and renewable resources. Non-renewable resources are finite – once they are used up they cannot be replaced because they take too long to form or regrow. These include the major fossil fuels formed over tens of thousands of years – coal, oil and natural gas.

FACT FILE: COAL

Status non-renewable fossil fuel
Description formed underground from decaying plant and animal matter
Lifespan over 200 years
% share of world energy use 30.0 (2014)
Main producers China, USA, Australia, India, Indonesia, Russia
Energy uses electricity, heating, coke
Advantages high world reserves; newer mines are highly mechanised
Disadvantages pollution: CO_2, the major greenhouse gas responsible for global warming; SO_2, the main gas responsible for acid rain; mining can be difficult and dangerous; opencast pits destroy land; heavy/bulky to transport

FACT FILE: OIL

Status non-renewable fossil fuel
Description formed underground from decaying animal and plant matter
Lifespan about 50 years
% share of world energy use 32.6 (2014)
Main producers USA, Saudi Arabia, Russia, China, Canada, UAE, Iran, Iraq
Energy uses electricity, petroleum, diesel, fuel oils, liquid petroleum gas, coke and many non-energy uses, e.g. plastics, medicines, fertilisers
Advantages variety of uses; fairly easy to transport; efficient; less pollution than coal
Disadvantages low reserves; some air pollution; danger of spills (especially at sea) and explosions

FACT FILE: NATURAL GAS

Status non-renewable fossil fuel
Description formed underground from decaying animal and plant matter; often found with oil
Lifespan 60 years
% share of world energy 23.7 (2014)
Main producers USA, Russia, Canada, Iran, Canada, Qatar, Norway
Energy uses electricity, cooking, heating
Advantages efficient; relatively clean – least polluting of the fossil fuels; easy to transport
Disadvantages explosions; some air pollution

SKILLS DECISION MAKING

ACTIVITY

Which of the non-renewable sources of energy do you think is best? Give your reasons.

There are two sources of energy that can be viewed as both non-renewable and renewable. If trees are planted for the specific production of fuelwood, then clearly the source is renewable. But if fuelwood is extracted by the felling of natural forests, then it is non-renewable. Nuclear energy involves the mining of non-renewable deposits of uranium to provide the necessary radioactivity. However, the radioactive material can be re-used and for this reason nuclear energy is renewable.

FACT FILE: FUELWOOD

Status non-renewable/renewable
Description trees, usually in a natural environment, but can be grown specifically for fuel
Lifespan variable within each country, but declining except where there is large-scale reforestation
% share of world energy use not known
Main producers of energy developing countries, especially in Africa and Asia
Energy uses heating, cooking (also used for building homes and fences)
Advantages easily available, collected daily by local people; free; replanting possible
Disadvantages trees are used up quickly; time-consuming – wood must be collected daily; deforestation leads to other problems (**soil erosion**, **desertification**); replanting cannot keep pace with consumption

FACT FILE: NUCLEAR ENERGY

Status classified by some as non-renewable because of reliance on uranium as a fuel; others regard it as renewable in that the nuclear fuel may be re-used
Description heavy metal (uranium) element found naturally in rock deposits
Lifespan unknown
% share of world energy consumption 4.4 (2014)
Main producers of energy USA, France, Japan, Russia, China, South Korea
Energy uses used in a chain reaction to produce heat for electricity

Advantages clean; fewer greenhouse gases; efficient, uses very small amounts of raw materials; small amounts of waste

Disadvantages dangers of radiation; high cost of building and decommissioning power stations; problems over disposal of waste; nuclear accidents such as Chernobyl have raised public fears

It is absolutely clear that the world needs to look to renewable sources of energy. It is not just that global stocks of the three fossil fuels will run out one day. Using them, and also fuelwood, harms the environment. Burning them certainly contributes to global warming. Nuclear energy, too, has environmental impacts, but of a rather different kind. The most notable impact is the risk of nuclear explosions and the safe disposal of nuclear waste. Despite the risks, many people think that much greater use will need to be made of nuclear energy in the future.

RENEWABLE SOURCES

A number of energy sources can be classed as renewable. These include:

- hydro (rivers, tides and waves)
- wind
- solar
- geothermal energy
- biomass.

Renewable resou rces are generally cleaner than non-renewable sources, but only produce around 15 per cent of the world's energy needs. They are often referred to as sources of **alternative energy**.

FACT FILE: HYDRO

Status renewable
Description good, regular supply of water needed; water held in a reservoir, channeled through pipes to a turbine
% share of world energy use 6.3 (2014)
Main producers China, Canada, Brazil, USA, Russia, Norway
Energy uses electricity
Advantages very clean; reservoirs/dams can also control flooding/provide water in times of shortage; often in remote, mountainous, sparsely populated areas
Disadvantages large areas of land flooded; silt trapped behind dam; lake silts up; visual pollution from pylons and dam

FACT FILE: GEOTHERMAL

Status renewable
Description boreholes can be drilled below ground to use the Earth's natural heat; cold water is pumped down, hot water or steam are channeled back
% share of world energy consumption less than 1 (2014)
Main producers Japan, New Zealand, Russia, Iceland, Hungary
Energy uses electricity, direct heating
Advantages many potential sites, but most are in volcanic areas at the moment
Disadvantages sulphuric gases; expensive to develop; very high temperature can create maintenance problems

FACT FILE: WIND

Status renewable
Description wind drives blades to turn turbines
% share of world energy less than 1
Main producers China, USA, Brazil, Canada, Germany
Energy uses electricity
Advantages very clean; no air pollution; small-scale and large-scale schemes possible; cheap to run
Disadvantages winds are unpredictable and not constant; visual and noise pollution in quiet, rural areas; many turbines are needed to produce sufficient energy

FACT FILE: TIDAL

Status renewable
Description tidal water drives turbines
% share of world energy consumption insignificant
Main producers France, Russia
Energy uses electricity
Advantages large schemes could produce a lot of electricity; clean; barrage can also protect coasts from erosion
Disadvantages very expensive to build; few suitable sites; disrupts coastal ecosystems and shipping

FACT FILE: SOLAR

Status renewable
Description solar panels or photovoltaic cells using sunlight
% share of world energy less than 1
Main producers USA, India
Energy uses direct heating, electricity
Advantages could be used in most parts of the world; unlimited supplies; clean; can be built into new buildings; efficient
Disadvantages expensive; needs sunlight and cloud/night mean that solar energy is reduced; large amounts of energy require technological development and costs in terms of photovoltaic cells

FACT FILE: BIOFUELS AND WASTE

Status renewable
Description fermented animal or plant waste or crops (e.g. sugar cane); refuse incineration
% share of world energy consumption 3.0 (2014)
Main producers Argentina, Brazil, Japan, Germany, Denmark, India
Energy uses ethanol, methane, electricity, heating
Advantages widely available, especially in LICs; uses waste products; can be used at a local level
Disadvantages can be expensive to set up; waste cannot be recycled; some pollution

SKILLS DECISION MAKING

ACTIVITY

Which of the renewable energy sources in your home country is likely to be most exploited?

Most sources of renewable energy seem attractive options. They directly focus on aspects of the environment that are inexhaustible. Most do so without any serious adverse impact on the environment. However, their big downside is that they cannot produce energy in the same huge quantities as the fossil fuels. Maybe one day new technologies will enable renewable sources to increase their contribution to the global energy supply. Bearing in mind the link between burning fossil fuels and global warming, the only option now is to make greater use of nuclear energy. There will be risks associated with this, but the world has no other choice – at least until the alternative renewable sources of energy contribute much more to the energy budget.

4.9 SUSTAINABLE ENERGY

In this part of the topic you should recall and make use of what you have learned from your research into different approaches to developing energy resources.

Two things are clear in today's world of energy, particularly as global demand increases and a larger strain is placed on global energy resources.

■ Energy must be used sparingly and with the utmost efficiency.
■ The non-renewable sources of energy are finite and must be conserved. We simply cannot afford to be wasteful; energy is a precious resource. Neither do we want the pollution caused by burning them.

Here we are repeating two points made earlier when discussing population pressure on resources in general (see page 112). We need to be sure that there will be enough energy to meet tomorrow's demands.

▶ **Figure 4.27: Making homes more energy efficient**

Renewable energy
Solar panels are one example of technology a homeowner can install to generate electricity for their home. This electricity can be used to run appliances, hot water services and energy-efficient devices such as under-floor heating.

Windows
Heat is often lost through windows. Double-glazing, heavy curtains and shutters can help reduce this loss.

Gas fires
Gas fires can be an economical method of heating smaller rooms. Doors should be kept closed so that the room retains the heat.

Insulation
Install insulation in your roofspace and inside wall cavities to keep a house cool in summer and warm in winter. Insulation can be made from fibreglass, cellulose, mineral wool and polystyrene. Polyurethane foam insulation can be useful for difficult to reach wall cavities.

Hot water service
Hot water services should be enclosed in a cupboard or small room to prevent heat loss. They should be upgraded to a newer model every 10 years to ensure maximum efficiency.

All of us as individuals have a responsibility to use energy more efficiently. There are many simple things that we can do in our everyday lives that will help save energy. A few examples are given here.

- Walk or cycle to and from school rather than rely on your parents to drive you.
- Homes in temperate latitudes lose an average of 50 per cent of their heat through the walls and loft spaces (Figure 4.27). Insulation is necessary to stop this. It may seem costly but in the long term it is more energy efficient. Curtains and blinds on all windows provide insulation in summer from the heat (reducing the need for air conditioning) and in winter help to keep the heat in (reducing the need for heating).
- Pack the empty spaces in the freezer and the refrigerator, either with ice trays or polystyrene. The more space that is taken up, the less energy it takes to cool or freeze. Do not run the dishwasher or the washing machine unless it is full.
- Many of us have electronic and electrical gadgets. Laptops, mobile phones, tablets, televisions and MP3 players all use energy. Putting your laptop and mobile phone into 'hibernate' for the night can save precious electricity. Instead of charging your mobile phone overnight, do so when you are still awake and unplug it as soon as it is done.

The following case studies look at the management of energy and its resources in three countries at different points along the development pathway.

SKILLS REASONING

ACTIVITY

Can you think of any more ways you might reduce your own energy consumption?

CASE STUDY: ENERGY RESOURCE MANAGEMENT IN QATAR

Qatar is one of a select group of countries enjoying energy security. One hundred years ago, it was one of the lowest income states in the Middle East. Thanks to the discovery of oil and natural gas, it is now one of the richest (Figure 4.28). Today it is the largest exporter of liquefied natural gas (LNG) in the world. The UK is one of the major importers. Qatar's exports of LNG, crude oil and refined petroleum products account for half its GDP. Thanks to these resources, the country has the world's highest per capita income.

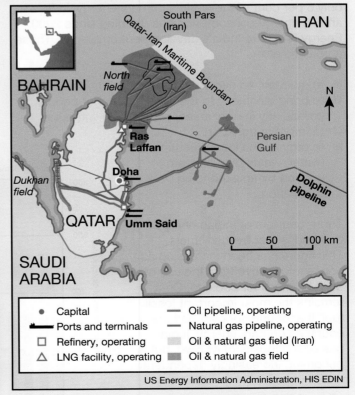

▲ Figure 4.29: Map of Qatar and its energy infrastructure

▲ Figure 4.28: Urban Qatar: a heavy consumer of energy

The physical geography and location of Qatar make it a major consumer of energy in the form of electricity. Its desert climate means it has a short mild, pleasant winter and a long, very hot and humid summer. In some months, the mean maximum temperatures exceed 40°C. As a consequence, there is a heavy demand for electricity from its population of nearly 2 million for air conditioning in the home, at work and in public buildings. Other consumers of electricity are the heavy industries (oil refining, fertilisers, steel and petrochemicals), the desalinisation plants that convert sea water into fresh water, and transport. Between 2000 and 2012, Qatar's electricity consumption grew from approximately 8.0 billion to 32.7 billion kilowatt-hours.

Responsibility for the management of the country's energy resources lies with the Emir and his Council of Ministers. While they are happy that Qatar should continue to use its natural gas to meet most of

ACTIVITY

Why does a small country such as Qatar have such a large demand for energy?

the country's energy needs, they are aware of its downsides. Although it is the cleanest-burning of the fossil fuels, it does produce carbon emissions. So, Qatar is beginning to look at another of its energy resources which, thanks to the climate, it has in abundance – solar power. Recently, it has announced plans to build a 1800 MW solar power station. By 2020, this is expected to account for around 15 per cent of the total electric power generation. It is doubtful whether Qatar needs to look for other renewable sources of energy. It has stated that it has no wish to turn to nuclear energy.

CASE STUDY: ENERGY RESOURCE MANAGEMENT IN INDIA

India is one of the leading emerging nations and has one of the fastest growing economies (Figure 4.30). It is home to 18 per cent of the world's population, but uses only 6 per cent of the world's primary energy. India's annual energy consumption has doubled since 2000. Energy is vital to India's development ambitions. Those ambitions are:

- to maintain an expanding economy
- to bring electricity to the large number of people, particularly in rural areas, who remain without it
- to improve transport
- to provide the infrastructure (water, waste disposal, etc.) needed by an expanding population.

Three-quarters of Indian energy demand is met by fossil fuels. This figure has been increasing as households move from the use of solid biomass for cooking. Coal remains the most important fossil fuel. It accounts for over half the country's primary energy. India has coal deposits of its own, but is becoming increasingly dependent

on imports. The same applies to oil and gas which meets over a third of the primary energy demand. Seventy per cent of India's electricity comes from the burning of fossil fuels.

Future management of India's energy supply seems unlikely to include any reduction in its dependence on fossil fuels. The country's primary focus seems to be on economic development, and more could be done to cut carbon emissions and global warming. However, there are potential sources of energy that India might eventually come to use. Hydro power has already

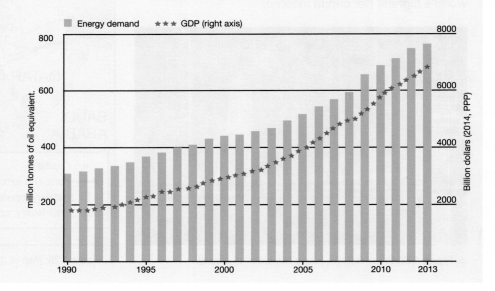

▶ Figure 4.30: India: primary energy demand and GDP (1990-2013)

been harnessed, particularly in the Himalayan area. It generates 20 per cent of the country's electricity. Nuclear power stations at present generate 5 per cent of the electricity. Wind power is beginning to be used in some of the coastal states. But there is one energy source that offers great potential and is as yet largely untapped – solar power. Figure 4.31 shows how great that potential is.

ACTIVITY

Can you think of one problem associated with using solar power in India?

ACTIVITY

Describe the main features of the distribution shown in Figure 4.31.

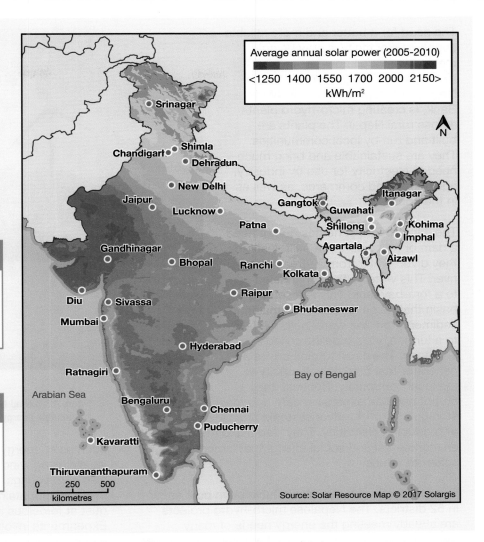

▶ Figure 4.31: Solar power potential in India

Average annual solar power (2005-2010)
<1250 1400 1550 1700 2000 2150>
kWh/m²

Source: Solar Resource Map © 2017 Solargis

CASE STUDY: ENERGY RESOURCE MANAGEMENT IN NEPAL

India's northern neighbour, the Himalayan kingdom of Nepal, is one of the lowest income countries in the world (Figure 4.32). The present demand for energy is relatively small. But it is growing as the country tries to develop and its people struggle for a better quality of life. Fuelwood was the traditional source of energy for heating and cooking. This led to widespread deforestation. Crop and animal waste (biomass) continues to be used as a fuel for cooking.

The ambition of the Nepalese government is to supply electricity to the population of 28 million. This is held back by the fact that Nepal has no significant deposits of coal, oil or natural gas of its own. Importing fossil fuels is made difficult and expensive because Nepal is a land-locked and mountainous country. There is an electric grid system covering part of the country and this could be extended to cover the more settled parts of Nepal.

▲ Figure 4.32: A remote rural community in Nepal

The need for a better and more reliable supply of electricity is now being met by what are known as micro-hydro plants. The government of Nepal, with support from the World Bank, is creating micro-hydro plants across rural Nepal. The plants are built and run by local communities. They are sustainable and bring much-needed electricity for use by industry, agriculture and commerce, as well as in the home.

The micro-hydro schemes do not need a dam or reservoir to be built. Instead, they divert water from a stream or river. This water is then channeled to a fore-bay tank. This is a settling basin that helps to remove damaging sediment from the water before it falls to a turbine, via a pipeline called a penstock. The small turbine drives a generator that provides electricity to the local community. Because they do not need an expensive dam and reservoir for water storage, run-of-the-river systems are a low-cost way of producing electricity. They also avoid the damaging environmental and social effects that larger HEP schemes cause.

Over 1000 micro-hydro plants have been built so far in 52 districts. The Nepalese micro-hydro projects are already meeting the energy needs of many rural communities. This shows that rural communities can also enjoy clean renewable energy.

Micro-hydro schemes are not the only sustainable sources of energy that can benefit Nepal. Attention has already turned to the possible use of other renewable

▲ Figure 4.33: Wind turbines and solar panels are two examples of sustainable energy.

sources of energy. Trial wind schemes in the Kathmandu valley, using wind generators, have shown promise but also some snags. The main problem is the wind itself. In highly mountainous countries, winds can suddenly gust at ferocious speeds that damage the windmills. Experiments involving the harnessing of solar power, either by solar panels (for heating) or by voltaic cells (for electricity), are also taking place. One obvious problem is the limited number of sunshine hours in this mountainous country. Another problem with both wind and solar power is the cost of the equipment and technology required to maintain it.

CHECK YOUR UNDERSTANDING

Why is Nepal so dependent on renewable sources of energy?

What conclusions can we draw from these case studies of three different countries?

- Despite their differences in levels of development, all three countries have energy needs. In Nepal, the need is to bring electricity to rural, often remote, areas. India needs all the energy it can get just to meet the needs of its rapid economic development. Qatar has to export its energy resources to maintain the country's economic prosperity. Much energy is also needed to keep the living and working environment in Qatar comfortable.
- When it comes to renewable sources of energy, there are both similarities and differences. All three have potential sources of renewable energy. However, Nepal is the only one to derive nearly all of its energy in this way – from hydro power. Qatar is beginning to make a small move in the direction of solar power. India already makes use of hydro power, but has no plans to become less dependent on fossil fuels.

CHAPTER QUESTIONS

END OF CHAPTER CHECKOUT

INTEGRATED SKILLS
During your work on the content of this chapter, you are expected to have practised in class the following:

- using numerical economic data to profile chosen countries
- interpreting photographs and newspaper articles
- using and interpreting line graphs showing changes in population and resources over time
- calculating carbon and **ecological footprints.**

SHORT RESPONSE

1 State **two** economic activities that belong to tertiary sector. [2]

2 Name **one** 'post-industrial' country. [1]

3 Identify **one** recyclable source of energy. [1]

4 Study Figure 4.26 (on page 118). Identify **two** countries that suffer from energy insecurity. [2]

LONGER RESPONSE

5 Define the term 'globalisation'. [4]

6 Suggest **two** reasons why the urban fringe is an attractive location for retailing. [4]

7 Assess the costs and benefits of using natural gas as a source of energy. [8]

8 Evaluate the reasons for using renewable sources of energy. [8]

EXAM-STYLE PRACTICE

1 a) Identify the meaning of the term 'GDP'. [1]
   A Grand diffusion process
   B Great development possibility
   C Gross domestic product
   D Good dairy produce

   b) Define the term 'non-renewable energy resource'. [1]

2 Identify the economic sector that includes mining. [1]
   A Primary
   B Secondary
   D Tertiary
   D Quaternary

3 Study Figure 4.21 (on page 114). Explain one way technology can help reduce overpopulation. [2]

4 a) State **one** factor affecting the location of manufacturing. [1]

   b) Explain **two** reasons why it is important for a country to have energy security. [2]

5 Study Figure 4.2 (on page 100). Suggest reasons for the decline of the secondary sector. [3]

6 For a named country, explain why its demand for energy is increasing. [6]

7 Study Figure 4.17 (on page 111). Analyse the reasons for the growth of informal employment. [8]

[Exam-style practice total 25 marks]

5 RURAL ENVIRONMENTS

LEARNING OBJECTIVES

By the end of this chapter, you should know:

■ What a biome is

■ The global distributions of biomes

■ That ecosystems provide goods and services

■ The ways in which people are exploiting ecosystems

■ The basic characteristics of a rural environment

■ The factors causing changes in rural environments in named countries

■ Ways of generating new income streams in rural environments

■ Strategies for making rural living more sustainable

■ The different groups involved in managing the challenges facing rural areas

This chapter starts by looking at the most important part of the rural environment – its ecosystems. Even in the countryside, ecosystems are exploited and changed by human activities, particularly by farming. It is this exploitation, together with the presence of people and their settlements, that convert natural environments into rural environments. Rural environments worldwide cannot escape the pressures of the modern world and as a result are undergoing great changes.

5.1 BIOMES AND THEIR GLOBAL DISTRIBUTIONS

The **biosphere** contains all the world's plant and animal life. That life or **biodiversity** may be broken down into major divisions known as biomes. There are 11 biomes altogether. You may be familiar with the names of some of them. Perhaps the best known is the tropical rainforest that occurs on either side of the equator (Figure 5.1). Less well known is the tundra found in the high latitudes near to the poles. The remaining nine biomes occur between these two global extremes.

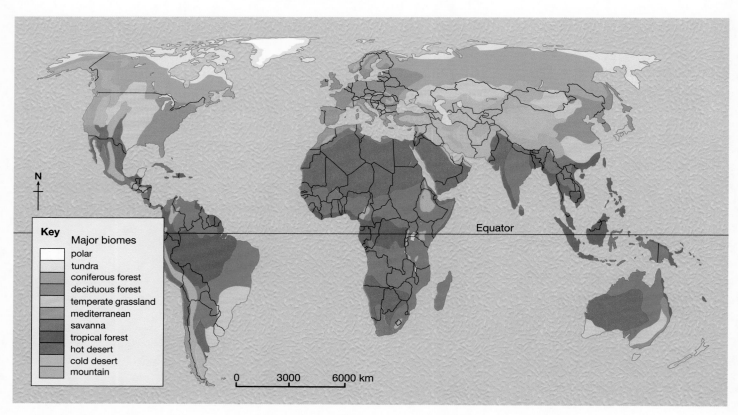

Key

Major biomes
- polar
- tundra
- coniferous forest
- deciduous forest
- temperate grassland
- mediterranean
- savanna
- tropical forest
- hot desert
- cold desert
- mountain

Equator

0   3000   6000 km

▲ Figure 5.1: The biomes of the world

The distributions of these 11 biomes are controlled by climate. Compare Figure 5.1 with Figure 5.2. Notice how both biomes and climate roughly change with latitude. Figure 5.3 shows how the two key factors of temperature and moisture affect biome distribution. Let us take a closer look at the nine largest biomes.

- Tropical rainforest – as the name indicates, this biome occurs where the climate is warm and humid, that is along the equator and close to it. Because of the constantly high temperatures and rainfall, vegetation grows more quickly than anywhere else on Earth. These conditions produce the greatest amount of living matter, referred to as primary productivity. The largest area of tropical rainforest occurs in the Amazon basin of South America.
- Savanna – this biome is also found along and close to the equator but where the climate is much drier. As a result, the primary productivity is lower. Instead of tall trees, there are scattered bushes and grasses. Savannah is most extensive in Africa.

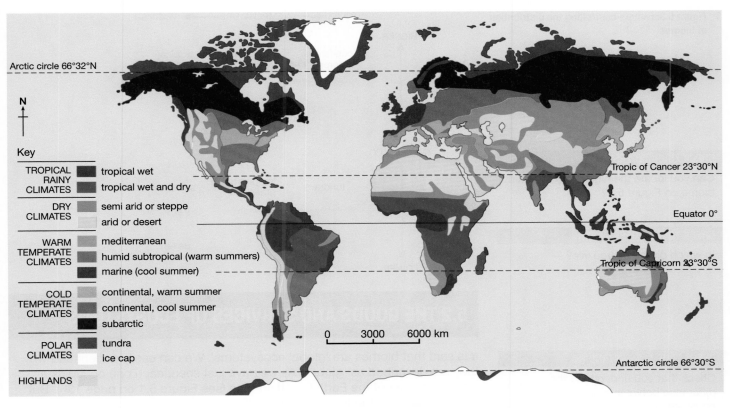

▲ Figure 5.2: The major climates of the world

Key

TROPICAL RAINY CLIMATES
- tropical wet
- tropical wet and dry

DRY CLIMATES
- semi arid or steppe
- arid or desert

WARM TEMPERATE CLIMATES
- mediterranean
- humid subtropical (warm summers)
- marine (cool summer)

COLD TEMPERATE CLIMATES
- continental, warm summer
- continental, cool summer
- subarctic

POLAR CLIMATES
- tundra
- ice cap

HIGHLANDS

■ Desert – the distinguishing feature of this biome is the general lack of vegetation due to lack of rainfall. As you will see in Figure 5.1, the biome extends over a wide range of latitude. Some of it occurs within the tropics, as in the Sahara Desert of Africa. Other parts of it are located outside the tropics, as in Asia. Because of this, the biome experiences a range of different temperature conditions.

■ Mediterranean – this biome is associated with a distinctive climate of winter rain and summer drought. The typical plants have adjusted to cope with the drought conditions. The characteristic vegetation is a mixture of small trees, low scrub and grassland.

■ Deciduous forest – because of low winter temperatures, the trees typical of this biome shed their leaves. There are three main areas: in North America, Europe and East Asia.

■ **Temperate grassland** – grasses dominate this biome because the climate is unsuitable for trees and shrubs. Large areas of temperate grassland occur in the dry continental interiors of North America (the Prairies) and Eurasia (the Steppes). We will take a closer look at this biome in Part 5.3.

■ Boreal (coniferous) forest – this biome occurs in high latitudes and stretches as a belt across North America and Eurasia. The leaves of the trees are needle-like to withstand the cold and loss of moisture.

■ Tundra – the long and bitterly cold winters, the short hours of winter sunshine, strong winds and the small amounts of precipitation are not favourable to plant growth. Vegetation is typically stunted and grows close to the ground. Grasses, mosses and lichens are most common. Dwarf trees are found in sheltered places. This biome stretches around the North Pole.

■ Highland – this biome occurs in the high mountains of the world. The climatic conditions are similar to those in the tundra. Highlands are very cold because temperatures decrease with altitude. This cold, plus strong winds, means that much of the precipitation falls as snow.

▶ **Figure 5.3: Factors controlling the distribution of biomes**

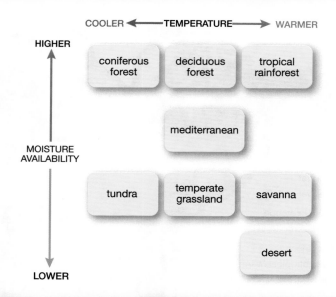

CHECK YOUR UNDERSTANDING

Can you locate the highland (mountain) biome on Figure 5.3?

CHECK YOUR UNDERSTANDING

In which biome do you live?

CHECK YOUR UNDERSTANDING

Check that you understand the difference between a biome and an ecosystem.

5.2 THE GOODS AND SERVICES OF ECOSYSTEMS

It is said that biomes are 'global ecosystems'. We can easily understand why they are described as 'global'. As we have just seen, each one occupies a significant area of the Earth's land surface (see Figure 5.1 on page 130). But what is meant by the term 'ecosystem'?

An ecosystem is a basic working unit or system of nature. It consists of living organisms (plants and animals) and their physical environment (sunlight, air, water, rock and soil). An ecosystem can be any size, ranging from a small pond to one of the 11 biomes, even to the Earth itself. The crucial thing is that the living organisms and physical environment are linked together. They are often reliant on each other for survival and maintain a balance that ensures each of them continues to exist.

Like all open systems, ecosystems involve inputs and outputs as well as internal flows of energy (Figure 5.4).

▶ **Figure 5.4: The inputs, stores and outputs of an ecosystem**

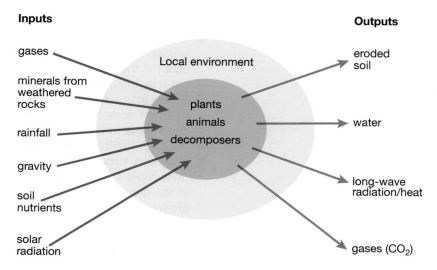

The greatest threat to the **well-being** and harmony of ecosystems comes from humans. The most obvious example is the long history of clearing ecosystems to create farmland. Many ecosystems offer resources and opportunities which people exploit. Today, these things are referred to as the goods and services of ecosystems. You might think of them as ecosystem outputs, along with the other outputs shown in Figure 5.4.

▶ Table 5.1: Goods and services of forest ecosystems

FOREST GOODS	FOREST SERVICES
Timber for building	Removal of air pollutants (carbon cycle)
Fuelwood	Emission of oxygen
Fodder for animals	Recycling of nutrients
Food (honey, fungi, fruit, nuts)	Recycling water
Medicinal plants	Maintaining biodiversity
Water	Recreation and leisure opportunities (tourism)

Goods are material things or products that can be taken directly from the ecosystem and put to use. Examples include timber and food. Table 5.1 illustrates some of the goods provided by forests. Services are the long-term benefits that people can gain from ecosystems. For example, mangroves provide coastal protection in tropical areas. Trees remove carbon dioxide from the air we breathe. Table 5.1 illustrates some of the services provided by forests. There are four broad types:

■ provisioning – supplying basic resources such as food and water
■ regulating – such as controlling climate (temperature and humidity) through the **carbon cycle** (Figure 5.5), and disease
■ supporting – such as cycling nutrients between biomass, litter and soil, and the water cycle
■ cultural – such as the recreational and spiritual benefits provided by an ecosystem.

▶ Figure 5.5: The carbon cycle

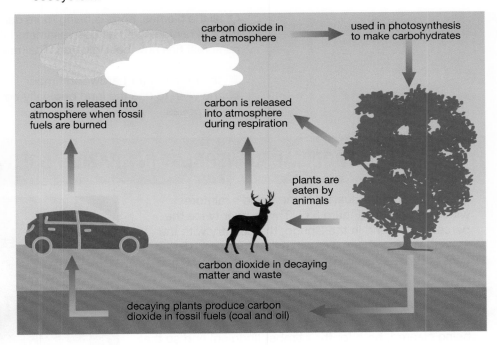

SKILLS ▶ SELF DIRECTION

ACTIVITY

Find out what the spiritual services of a forest might be.

SKILLS ▶ PROBLEM SOLVING

ACTIVITY

Complete a table showing what you think are the goods and services of grassland ecosystems.

Perhaps the forest service most talked about today is the part it plays in the carbon cycle (Figure 5.5). Forests are carbon stores and extract carbon dioxide from the atmosphere. If there were no forests, the global climate would become hotter and hotter to the point that plant and animal life would no longer be possible.

5.3 THE IMPACTS OF ECOSYSTEM EXPLOITATION

Human exploitation of the goods and services of ecosystems is an important part of the way of life in rural areas. Clearly, people benefit from them. Unfortunately, most often the ecosystems do not. The exploitation of these resources (goods and services) has negative impacts. Ecosystems are changed so as to threaten their future well-being, and even their continued existence. Just think what damage and destruction deforestation can cause.

On a global scale, however, the greatest single human impact on ecosystems has been through farming and people's need for food. Here we should distinguish between different types of farming.

ACTIVITY

Give examples of how ecosystems are changed by farming.

- **Subsistence farming** or commercial farming – except in areas of acute population pressure (overpopulation), the impact of subsistence farming on ecosystems may be just about tolerable. However, whole ecosystems are obliterated by commercial farming – because of its scale, its modern and powerful technologies (mechanisation, irrigation, GM crops, disease and pest controls), and the pressure to maximise profits.
- **Arable farming** or **pastoral** (livestock) **farming** – both types involve ecosystem clearance and change. In general, however, a hectare of arable farmland can feed more people than if it were used for rearing livestock.
- Intensive or **extensive farming** – both can be highly destructive. Intensive farming (such as market gardening) involves such a concentration of inputs that 'artificial' ecosystems are created, for example within **glasshouses** or poly-tunnels. Extensive farming uses large areas (for example, hill sheep farming and prairie farming in the USA).

Of all the biomes, none is being more damaged by human activities than the tropical rainforest. It has been estimated that about half the global stock of tropical rainforest has now been cleared. The main cause of deforestation has been:

- the demand for hardwood timber
- the wish to use the land originally covered by trees for totally different purposes.

CASE STUDY: TROPICAL RAINFOREST IN BRAZIL – USES AND IMPACTS

Brazil has about a quarter of the world's rainforest within its borders (Figure 5.6). This is over twice the percentage found in the Congo, the country with the second largest share. It is estimated that over 15 per cent of Brazil's rainforest was lost between 1970 and 2015.

USES BEING MADE OF THE RAINFOREST
In Brazil, huge areas of forest have been and still are being cleared, both for their timber (logging) and so that the cleared land can be used for:

- pasture for cattle rearing
- arable land for growing biofuels
- building land for settlements and roads.

▲ Figure 5.6: The vastness of the Brazilian rainforest

Of course, the trees that are felled as the land is cleared are sold for their timber. Hardwoods fetch high prices and this makes deforestation profitable. But it is how the cleared land is used that is the main driver of deforestation. This explains why logging is shown in Figure 5.7 as being directly responsible for such a small percentage of the deforestation.

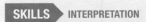
SKILLS ▷ INTERPRETATION

ACTIVITY

Check on Figure 5.1 the location and extent of the tropical rainforest biome.

SKILLS ▷ INTERPRETATION (STATISTICAL)

ACTIVITY

Why is logging in Figure 5.7 shown as such a small cause of deforestation?

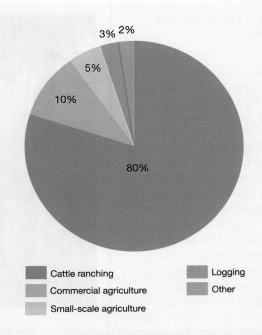

▲ Figure 5.7: The causes of deforestation in the Brazilian rainforest

Figure 5.7 shows that livestock rearing, a type of extensive farming, accounts for 80 per cent of the deforestation. The cleared land is sown with grasses to create pasture. However, the land cannot be used for long as the quality of the pasture declines rapidly. The cattle farmers then have to move on and destroy more rainforest to create new cattle pastures.

About 10 per cent of the forest is being cleared to make way for huge plantations, where crops such as soybeans, palm oil and sugar cane are grown for conversion into biofuels. Other crops include bananas, pineapples, tea and coffee. As with cattle ranching, the soil will not sustain crops for long. After a few years, more rainforest is cut down for more plantations.

Roads are needed to bring farm equipment and transport products to market. But building new roads means cutting through the rainforest. The trouble is that once the roads have been built, the forest becomes accessible to other commercial activities; examples are the mining of gold and bauxite, and the damming of the tributaries of the River Amazon to create reservoirs for the generation of electricity.

Other goods and services of the rainforest are being exploited, as well as the land it occupies. We have already mentioned exploiting the timber (logging). This

does not necessarily mean **clear felling**. **Selective logging** and **agro-forestry** are more sustainable ways of exploiting timber. There are other forest resources which are harvested; these include medicinal plants, fruits and nuts.

It is important to remember that indigenous people have lived in the rainforest for centuries. They have a traditional way of life that is closely geared to the resources of the forest. Over a period of many years they have:

- cut wood to burn as a fuel
- used timber to build their dwellings
- harvested fruits and nuts
- used plants for the treatment of illnesses
- cleared small areas of forest for food production.

SKILLS ▷ DECISION MAKING

ACTIVITY

Discuss, in groups, the claim that roads are the tropical rainforest's worst enemy.

▶ Figure 5.8: A treehouse in the rainforest

IMPACTS

The clearance of land for subsistence farming by indigenous people has done very little lasting damage to the forest. When the soil becomes exhausted, they move on and clear another area. This **shifting cultivation** means that, once abandoned, the forest is able to regenerate. Unfortunately, this is not the case with most of the other users of the rainforest land. Deforestation has a whole range of disastrous consequences including loss of biodiversity, land degradation and erosion, global warming and localised climate change, displacement of indigenous people and pollution of rivers.

INDIGENOUS PEOPLE

Some would argue that the exploitation of the tropical rainforest is helping Brazil's economic development. Therefore, everyone in Brazil is benefiting. But the claim is false. Just think what is happening to the indigenous people. Their traditional way of life is closely linked to the resources of the rainforest. For centuries now, they have been gradually forced out of their homelands by:

- logging
- mining
- the creation of ranches, plantations and reservoirs
- the construction of roads and settlements for incomers.

Many of these displaced people have ended up in towns and cities. Few have adjusted to this very different environment. Addiction to drugs and alcohol has been common. Many have died young. There are now about 240 tribes left in the rainforest compared with 330 in 1900. With the loss of these tribes have gone centuries of detailed knowledge of how to live sustainably in the rainforest.

To add to these human costs of deforestation, there have been at least five negative physical impacts.

SOIL EROSION

As soon as any part of the forest cover is cleared, the nutrient cycle is broken (Figure 5.09b). The thin topsoil

SKILLS ▷ DECISION MAKING

ACTIVITY

Imagine you are from one of the indigenous tribes and have to meet a government official. What arguments would you use to support the case for making your tribal lands into a protected area?

is quickly removed by heavy rainfall. Bare slopes are particularly prone to soil erosion. Once the topsoil has been removed, there is little hope of anything growing again. Soil erosion also leads to the silting up of rivers.

Even where the soil is protected, it quickly loses what little fertility it had when covered by trees. Grazing and plantations do little or nothing to keep the soil fertile. It is this decline in soil fertility that leads to pastures and plantations being abandoned. So, more areas of rainforest are cleared.

RIVER POLLUTION

Gold mining not only causes deforestation. The mercury used to separate the gold is allowed to enter the rivers. Fish are poisoned as well as people living in nearby towns. Rivers are also being polluted by soil erosion.

LOCAL CLIMATE CHANGE

Deforestation disrupts the water cycle (Figure 5.09a). With the felling of trees, evapotranspiration is reduced and so is the return of moisture to the atmosphere. The local climate becomes drier. The recycling of

▶ Figure 5.9: Rainforest cycles of:
   a) water, and b) nutrients

Rainforest cycle of water

heavy convectional rain

evaporation

trees protect ground from most rain

trees absorb water

some rain gets to the ground

a

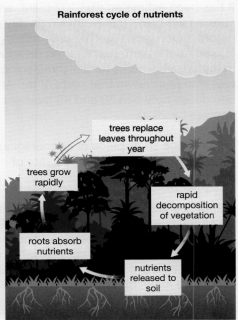

Rainforest cycle of nutrients

trees replace leaves throughout year

trees grow rapidly

rapid decomposition of vegetation

roots absorb nutrients

nutrients released to soil

b

ACTIVITY

What else might be causing river pollution?

water is like a cooling system. So, once the recycling is reduced, the local climate becomes warmer. The combination of increasing dryness and rising temperatures is not a good one, either for people or activities such as commercial farming that follow the clearance of the rainforest.

GLOBAL WARMING

The rainforest is even more significant at a global level. The tree canopy absorbs carbon dioxide in the atmosphere. This stops, of course, as soon as the trees are felled and so more carbon dioxide remains in the air. The clearance of rainforest usually makes use of fire. This means that carbon stored in the wood returns to the atmosphere – adding even more carbon dioxide to the atmosphere. Carbon dioxide allows the Sun's rays to come through the atmosphere. However, it does not allow heat from the Earth's surface to escape from the atmosphere. This is called the greenhouse effect. It is a cause of the global warming that is threatening the survival of the human race – not just the people of the tropical rainforest.

LOSS OF BIODIVERSITY

One of the remarkable features of the tropical rainforest is its high level of biodiversity. Indeed, the level is higher than in any other biome. More than two-thirds of the world's plant species are found in the tropical rainforests. They also contain about half of the world's known animal species, ranging from birds and mammals to reptiles and insects.

If the rainforest is cleared, two things will happen:

■ global biodiversity will be much reduced
■ individual species will become endangered and then possibly extinct.

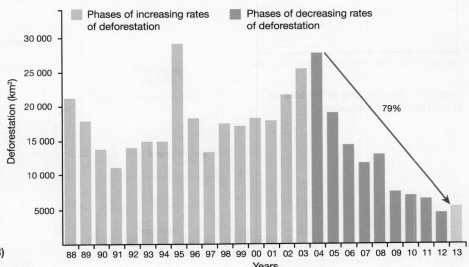

▶ Figure 5.10: Deforestation rates (1988–2013)

It has been estimated that 137 plant, animal and insect species are being lost every single day due to rainforest deforestation globally. That amounts to 50 000 species a year. As the rainforest species disappear, so do many possible cures for life-threatening diseases. New research has shown that parts of the Amazon rainforest could lose between 30 and 45 per cent of their main species by 2030.

There is one small piece of encouraging news from the Brazilian rainforest. The annual rate of deforestation has decreased since 2004 (Figure 5.10).

Disputes arise because people have conflicting views about the use and future of rainforest. There is clearly a tension between developers and conservationists. The indigenous people are in conflict with those who wish to exploit the rainforest in any large-scale way.
The future looks bleak for them and the rainforest.

ACTIVITY

Write an account summarising the information given in Figure 5.10.

SKILLS   DECISION MAKING

ACTIVITY

Of all the local impacts, which do you think is most serious? Give your reasons.

5.4 CHARACTERISTICS OF RURAL ENVIRONMENTS

ACTIVITY

See Part 6.1 (page 158) for more information about urbanisation.

In this part of the topic you should recall and make use of what you have learned from your research into the changing use of rural environments.

All countries today show the same major division between rural environments and urban environments. The process of urbanisation (becoming more urban) now affects all parts of the world. As a result, rural areas are becoming less populated. They are being depleted by the spread of urban areas. Overall, the world's urban population is growing more rapidly than the rural population (Figure 5.11).

So, what are the characteristics of rural environments? What makes them attractive to some people and unattractive to others? The main distinguishing components are shown in Figure 5.12.

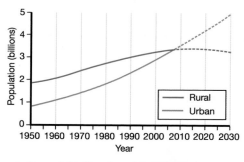

▲ Figure 5.11: The changing global balance of rural and urban populations

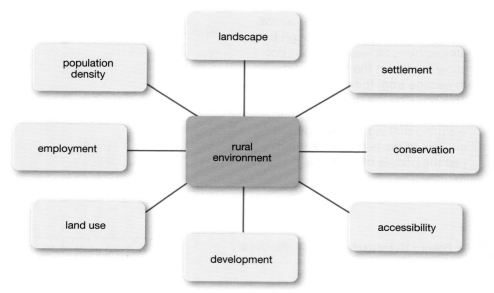

▶ Figure 5.12: Distinctive components of rural environments

▲ Figure 5.13: Three rural environments from around the world: a) Switzerland, b) Chile, c) Sri Lanka

SKILLS INTERPRETATION

ACTIVITY

Write short descriptions of each of the rural landscapes shown in Figure 5.13.

SKILLS DECISION MAKING

ACTIVITY

Think about a rural area that you know well. Which of the eight components (Figure 5.12) contributes most to its rural character?

■ **Landscape** – for many people, the term rural environment suggests an image of a landscape with scattered settlements separated by open spaces – mainly green spaces given over to farming. The landscape is created by the physical factors of geology, relief and climate. These factors strongly influence what people can do for a living. What type of farming is possible? Will the scenery attract tourists? Will it be a pleasant place in which to live?

■ **Population density** – rural areas have relatively low population densities compared with the high densities of urban environments. However, in some parts of the world rural settlements can be both large and very compact. Population densities within them can be as high as in many towns and cities. High mean densities in rural areas can occur in rural countries such as Bangladesh.

■ **Settlement** – the settlement pattern is typically made up of villages, hamlets and farmsteads. Settlements are separated from each other by unsettled tracts that are either farmed or left in a fairly natural state.

■ **Employment** – most jobs in rural areas belong to the primary sector. By far the most important employer is farming and the production of food. Other forms of rural employment include fishing, forestry and sometimes the working of minerals. In general, these jobs are poorly paid. It is because of this that so many people leave the countryside in search of better paid jobs in towns and cities.

■ **Land use** – while farming is the biggest use of land in many parts of the world, there are parts of the rural environment where other activities are important. Traditionally, forestry was one such activity in areas with wooded ecosystems. Much mining and quarrying continue to take place in largely rural environments. However, there are two significant new activities – leisure and tourism. The urban populations of the world look increasingly to rural environments; not just for food, but also as places in which to relax and take a break from the routine of work.

■ **Accessibility** – it is claimed that one of the reasons for the lack of development in rural areas is their lack of accessibility. Certainly, there are many remote rural areas in the world. This is particularly the case in the more mountainous, arid and forested regions. However, the situation is changing. The improvement of rural roads, in particular, is allowing for more leisure and tourism in the rural environment.

■ **Development** – population pressure in many parts of today's world means that more use needs to be made of rural spaces. In some countries, this involves producing more food. In others, new activities are encouraged which, in turn, develop more jobs and reduce the dependence on agriculture as a means of making a living.

■ **Conservation** – a change is taking place in some rural environments. The rise of leisure and tourism is one example of a different way in which rural environments are valued. Another is that people are realising that the conservation of rural environments and their wildlife is important for maintaining the world's biodiversity. Conservation is also seen to be vital in terms of helping to reduce global warming as trees recycle carbon dioxide into oxygen.

Rural environments are valuable and have much to offer modern society. In these days of global urbanisation, it is too easy to think of rural environments as being less important than the booming urban areas of the world.

5.5 RURAL CHANGE IN THE UK

In this part of the topic you should recall and make use of what you have learned from your research into the changing use of rural environments.

CASE STUDY: CHANGING RURAL AREAS OF THE UK

The balance of the components shown in Figure 5.12 (page 138) is constantly changing in response to outside influences. In the UK, as in other developed countries, the major influence was the urbanisation process. The UK is now in the late stages of urbanisation. The main movement of people and activities is now outwards from urban centres, rather than towards them.

The impacts of this recent change in the direction of urbanisation vary from place to place within the UK. The key factors are distance and accessibility to large cities. Four broad types of rural area may be recognised in the UK and in many other developed countries (Figure 5.14).

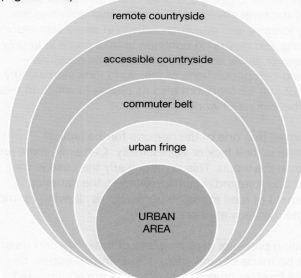

▲ Figure 5.14: Four types of rural area

We will now look at the changes associated with each of these.

URBAN FRINGE

This is the rural space on the edge of the built-up area of towns and cities (Figure 5.15). It is rural space that is being taken over by the outward spread of the built-up area. The main process of change here

is suburbanisation. This involves the building of new homes, most often in the form of large housing estates. These estates become occupied by either rural-urban migrants or people who choose to move out of the older parts of the built-up area. The latter are attracted by the new housing and nearness to the open countryside. However, the occupiers of these new homes need more than just a house. They also need services, such as shops, schools and medical centres. Other activities will also be attracted by the increasing number of people living in the new suburbs (see Part 4.2, page 102). Industrial estates, business and science parks will be drawn by the availability of labour nearby, but also by better accessibility (provided by ring roads and bypasses), cheaper land, and the attractive setting provided by the remaining countryside.

There is not enough space in the urban fringe to meet the demand for new homes and various kinds of 'parks'. This has caused planners to act to protect the remaining greenfield sites and to create green belts around cities (Figure 5.16). Within these belts, there are strict controls that allow few, if any, new developments. But the demand for housing in the UK has become so great that some green belt land has been released for development.

▲ Figure 5.15: The new urban fringe landscape

ACTIVITY

What do you think are the consequences of creating a green belt around a town or city?

▶ **Figure 5.16: Urban green belts in England and Wales**

THE COMMUTER BELT

The commuter belt lies just outside the urban fringe (Figure 5.14). Change here is largely due to the arrival of commuters. These are mainly people who work in a nearby city or town and use its services. They are tempted to move here by:

- the attraction of cheaper and more spacious housing
- the availability of fast transport to the place of work; most jobs are located either in a city centre or in new industrial estates and business parks in the urban fringe
- the opinion that the commuter belt offers a better quality of life and is a better environment in which to bring up a family.

Many people feel that the time and money spent on commuting is worth it. Figure 5.17 shows how far commuters in south-east England are prepared to travel each working day.

SKILLS ▶ PROBLEM SOLVING

ACTIVITY

On Figure 5.17, measure the straight-line distances being covered by some of London's long-distance commuters.

Key
- Urban areas
- Green belts

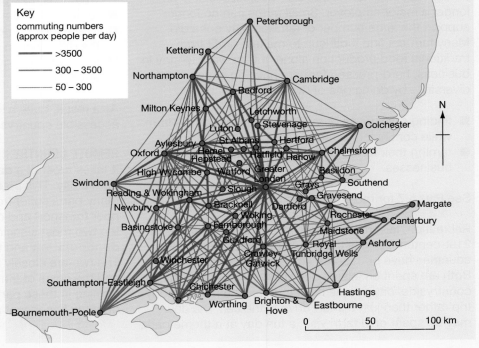

Key
commuting numbers
(approx people per day)
- >3500
- 300 – 3500
- 50 – 300

▶ **Figure 5.17: Commuting in south-east England**

▲ Figure 5.18: Using the accessible countryside for recreation and leisure

The impact of commuting has changed the character of rural villages and hamlets. These detached suburbs are also called 'dormitories'. New housing estates are built on their edges. Older housing is modernised and often extended. The old village shops are replaced by small supermarkets. Rural crafts and the links with farming disappear. Communities often become divided into two with the newcomers gradually overwhelming the original inhabitants.

ACCESSIBLE COUNTRYSIDE

Accessible countryside lies beyond the commuter belt, but is a day trip's reach from towns and cities. It is still very much a rural area. Three important changes are taking place in these areas. These are changes in farming, recreation and retirement migration.

Farming: Although farming remains a significant land use in the accessible countryside, its character is changing. Because of mechanisation and the amalgamation of farms into agri-businesses, it is no longer a major employer. Small farms can no longer support the large number of families they once did. Many farmers find it difficult to make a profit from traditional food production alone. If they want to stay in business, hard-pressed farmers have no choice but to diversify – by doing one of two things:

■ finding other ways of making money out of the farm, while continuing to farm
■ turning their farms into completely different businesses.

See Part 5.7 for more about both options.

Recreation, leisure and tourism: One of the features of 21st-century living is that many people in developed countries have both spare time and disposable income. Both are spent on leisure activities. In the accessible countryside, city people have to think in terms of a day trip rather than just a short drive in the car. The day may be spent on a farm visit, a fun day at a theme park, some birdwatching around a nature reserve or pony trekking (Figure 5.18).

The growth of recreation, leisure and tourism in the accessible countryside is not evenly spread. It is focused on what are called honeypots – places that offer something that attracts large numbers of visitors. It might be especially attractive scenery, picturesque settlements or some well-known historic connection. But recreation, leisure and tourism are putting great pressure on these popular locations. The most obvious symptoms are crowds and traffic congestion.

Retirement migration: People in developed countries are living longer. Most people can expect to enjoy 10 or more years of retirement. As a result, more people are moving home once they have retired. They are doing this for a number of reasons:

■ it is no longer necessary for them to live close to what was their place of work
■ to downsize to a smaller home
■ to sell their home for something cheaper and to use the difference in price as a sort of pension
■ to move into a quieter, calmer and more attractive environment.

As a result, quite a lot of retirement migration is to the accessible countryside.

REMOTE COUNTRYSIDE

It takes most of a day to reach remote countryside from a city. This countryside is almost totally rural. Many parts suffer from rural isolation due mainly to poor transport and communication links. Probably the two most significant changes in these areas are depopulation and the dramatic decline in farming. The two are closely linked. Remoteness from markets and the costs of transport are two factors making farming in these parts uneconomic. The abandonment of farms has meant a serious loss of jobs.

ACTIVITY

Compare the challenges facing the UK's accessible and remote countryside.

ACTIVITY

In class, debate the claim that rural change in the UK is a result of urbanisation.

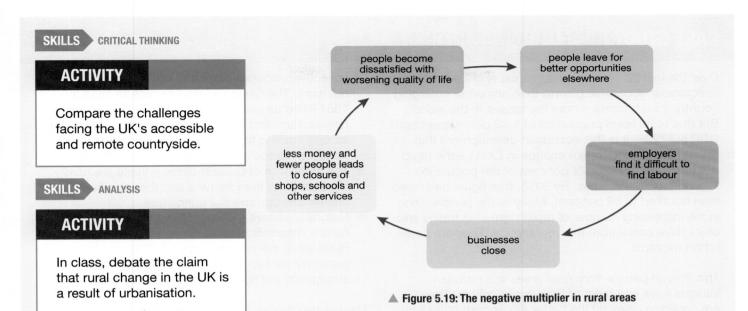

▲ Figure 5.19: The negative multiplier in rural areas

For this reason, several generations of young people have been forced to leave the remote countryside in search of work in towns and cities. They have also been persuaded to move because of the declining quality of life. This decline in farming and loss of population has a negative multiplier effect and leads to a spiral of decline (Figure 5.19). The fact that a growing number of wealthy people are buying second homes does not really do anything to stop the negative multiplier.

However, it is not all bad news in the remote countryside. There are two positive changes. Fortunately, within the remote countryside, tracts of country have been designated as national parks. This has occurred in recognition that they offer something special. It may be spectacular scenery, wilderness or biodiversity. National parks and other protected areas attract holidaymakers and tourism is now offering all sorts of employment opportunities – from jobs in hotels and restaurants to jobs as guides and park rangers.

The downward spiral is also being slowed down by the process of **counter-urbanisation**. The loss of population from remote countryside is now beginning to be compensated by an inward movement of newcomers. Thanks to the broadband revolution, people tired of urban living and tempted by a rural setting are able to work from home in these remote parts.

Clearly, change in the rural areas of the UK beyond the urban fringe has led to a number of new challenges. These include:

■ making changes in farming
■ finding new sources of rural income
■ improving the quality of rural life
■ protecting and upgrading the rural environment.

Dealing with these challenges will be covered in Parts 5.7, 5.8 and 5.9.

5.6 RURAL CHANGE IN CHINA AND KENYA

In this part of the topic you should recall and make use of what you have learned from your research into the changing use of rural environments.

Rural change, particularly in emerging and developing countries, is strongly influenced by four factors:

■ economic development
■ technological development
■ population growth
■ rural-urban migration.

The following case studies illustrate an important point. These four factors are more likely to result in problems rather than benefits for rural areas.

CASE STUDY: RURAL CHANGE IN CHINA

Over the last 20 years, China has experienced more spectacular economic growth than any other emerging country. Its economy is now the largest in the world. But due to its huge population of 1.35 billion, per capita GDP is $7295. It is this economic development that lies behind another major change in China – the rapid urbanisation. In 1990, 21 per cent of the population lived in towns and cities. By 2015, that figure had more than doubled to 56 per cent. Many of the people living in the increasing number of rapidly growing towns and cities have come from the countryside. They are rural-urban migrants.

This flow of people from rural areas is a problem. Villages have lost most of their young adults. They are drawn to cities by the better job opportunities and quality of life, as well as by the more exciting urban lifestyle. As a result, farming has lost much of its labour force and its output has declined. The country is now faced with a major problem.

Urbanisation has worsened the food supply situation in two other ways:

- the growing urban population needs feeding – but where is the food to come from?
- farmland is being lost as the built-up areas spread rapidly outwards.

It is unfortunate that, even before the exodus of people to the cities, Chinese agriculture was renowned for its low productivity, due to the following factors:

- Little mechanisation.
- Limited availability of fertilisers and pesticides.
- Changing political ideas about what was best for food production. After 1949, when the present Communist China came into being, all land became state-owned. Farmers were organised into collectives, cooperatives and communes. These did not work. The Great Chinese Famine of 1958 to 1961 killed as many as 40 million people. After the famine, farmland was again divided up and shared among farming families. But the resulting farm holdings are too small for commercial farming.

- The smallness of present farms – these are hardly big enough to feed farmers and their families, and unsuited to commercial agriculture.
- Natural disasters – China suffers from three types of natural disasters: floods, droughts and earthquakes. Rural areas are usually badly hit. Their plight is made worse by the fact that urban areas often receive emergency aid and help with recovery first.

The net result of all these factors is that China is having trouble feeding its huge population of 1.35 billion people. It has to import more and more food. At the same time, rural people are missing out on the benefits of economic development enjoyed by the urban population. They live in poor housing, have access to few services and suffer a poor quality of life.

Change in rural China is caused by economic development and urbanisation. But the change appears to be for the worse. It is trapping China's rural areas in a downward spiral of poverty and deprivation.

SKILLS SELF DIRECTION

ACTIVITY

Find out what impact the Cultural Revolution had in the countryside of China.

DID YOU KNOW?

China is one of the most hazardous countries in the world. It has suffered two of the deadliest earthquakes in human history. Its famines have in recent times killed tens of millions of people. Its floods have been among the most damaging ever recorded.

▲ Figure 5.20: Chinese farming still needs labour.

▲ Figure 5.21: Chairman Mao Zedong bringing the Cultural Revolution to the countryside

CASE STUDY: **RURAL CHANGE IN KENYA**

Compared to China, Kenya is further behind on the development pathway. Per capita GDP is only 20 per cent of that in China. Although only a quarter of the 45 million population lives in urban areas, the rate of urbanisation has been increasing at an annual rate of just over 4 per cent. As in China, there are strong flows of rural-urban migration. So, we can expect that a major cause of change in Kenya's rural areas will be the loss of population resulting from urbanisation.

You should be familiar with the attractions that pull people towards towns and cities. It is usually males who move. This has one immediate consequence. Those members of the family who are left behind are less able to carry out the heavy work of subsistence farming. Often **malnutrition** and sometimes starvation follow.

What are the rural push factors?

■ The low productivity of farming – partly due to harshness of the physical environment in many areas, and partly due to poor farming techniques.
■ Poor access to basic education and healthcare services. While **literacy** rates in Kenya's towns and cities are among the best in Africa, the rate in rural areas is poor.
■ Corruption – it has been claimed that the better farmland is taken from its owners and given to officials or company bosses as bribes.
■ Increasing frequency of droughts has a devastating impact on the basic food supply (Figure 5.22). Less frequent floods also interrupt food supply.
■ The lack of help to deal with the human hazard of HIV/AIDS. Treatments are available in the major towns and cities.
■ Finally, but by no means least, the spread of commercial agriculture which has taken over some of the best farmland (see Part 5.8).

Even before rural-urban migration set in, there were problems in many rural parts of Kenya. The recent changes caused by urbanisation have simply added to those problems and led to increasing poverty. In this respect, the cases of Kenya and China are remarkably similar.

SKILLS ▸ DECISION MAKING

ACTIVITY

Imagine you are living in rural Kenya. What arguments would persuade you to stay where you are? What arguments would persuade you to move to Nairobi?

SKILLS ▸ INTERPRETATION

ACTIVITY

Look at Figure 5.22. What observations would you make about subsistence farming in this part of Kenya?

▲ **Figure 5.22: The struggle of subsistence farming in semi-arid areas**

ACTIVITY

Look at the pie charts in Figure 4.14 (page 108) to see what contribution farming makes to the UK economy.

5.7 THE DIVERSIFICATION OF FARMING AND FARMS

The challenge of how to make the rural areas in many developed countries more sustainable starts with the following questions.

■ Is it possible to raise the productivity and profitability of farming?
■ If 'yes', then how?
■ If 'no', then what might farmers do to generate new streams or sources of income?

This particular challenge is best explored by looking again at the UK.

CASE STUDY: FARM DIVERSIFICATION IN THE UK

Farming in the UK today is very different to what it was 200 years ago when it was a major part of the country's economy. Today, it accounts for about 1 per cent of GDP and 1 per cent of employment (Figure 5.23). Statistics also show that agricultural output has increased considerably. This is due to:

■ mechanisation
■ making farms larger
■ specialising in particular lines of farm produce
■ taking advantage of many advances in farm technology.

Advances include using more effective fertilisers, pesticides and herbicides. Crop and livestock strains have been improved, thanks to breeding programmes and **genetic engineering**.

Commercial pressures from the major food retailers, such as Tesco and Sainsbury, are also persuading farmers to become more efficient and productive. However, many UK farmers may be wondering if there is a limit to how much agricultural productivity can be raised. The rate of increase in agricultural productivity in the UK is definitely slowing down.

One possible way of raising agricultural output still further is to grow more **genetically-modified (GM) crops**. This is highly controversial (Table 5.2). Some argue that it would not raise production by much as disease and pests are already well controlled by other methods. Others are worried about the impact of GM crops on the environment, particularly wildlife.

The fall in the prices paid to farmers for their crops and livestock might suggest that the UK farmers are producing too much food. Certainly, a fair amount of farmland has been taken out of production. But is UK farming really overproducing? The reason for asking this question is quite simple. The UK now only produces 60 per cent of the food it needs. It relies on other countries

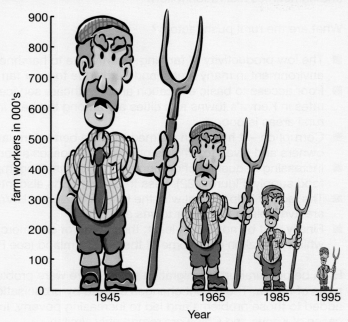

▲ Figure 5.23: The declining labour needs of UK farms

ARGUMENTS FOR	ARGUMENTS AGAINST
Higher crop yields	Possible contamination of other plants
Cheaper food	Possible adverse impact on human health
Better quality food	Expensive feedstock
Less use of herbicides	Some crop plants do not produce seeds

▲ Table 5.2: The GM debate

ACTIVITY

Give some examples of exotic supermarket foods. Where have those foods come from?

▲ Figure 5.24: Some different ways of raising the income of a UK farm

to provide the remaining percentage. Surely, that would not happen if UK farming was really overproducing?

This strange situation of the UK importing an increasing amount of its food is the result of two common beliefs among the public:

■ that cheap food is a 'must'
■ that 'seasonal' foods, such as strawberries, beans and apples, should be on the supermarket shelves throughout the year, and not just during the UK growing season.

There is also a growing market for produce that cannot easily be grown in the UK due to its climate.

As a result, the UK's imports of food are made up mainly of:

■ food that can be grown more cheaply outside the UK
■ food that can only be grown in the UK during the summer months.

There are two ways that UK farmers can earn more money: a) by raising more income from farming activities, and b) by putting their farm – its buildings and land – to non-farming uses.

Five different ways of raising more income from farming activities are currently being tried:

■ Farm shops – these involve the farmer selling directly to customers at the farm gate. At least this cuts out the need for a wholesaler and retailer. Such shops often include a cafe or restaurant which provides another outlet for the farm's produce. Closely allied to the idea of the farm shop are farmers' markets. Groups of farmers get together regularly to sell directly to the public at some accessible location.
■ Organic farming – for health reasons, some people are concerned about the chemicals being used in the growing of crops and the rearing of livestock. Organic farming does without those chemicals, but is less productive. So, the food does cost more. However, demand for organic produce is increasing in the UK and some customers are prepared to pay a little more for it. It is even better for farmers if the organic food is bought at a farm shop.
■ Pick-your-own – allowing customers to pick their own produce such as soft fruit and vegetables saves on labour costs as well as the costs of transport.

■ Novel products – this is the term used to describe a range of unusual crops and livestock that can be grown in the UK for specialist markets. They can range from lavender and edible flowers to angora wool and ostrich meat.
■ B & B – a short stay on a working farm in a pleasant rural location is popular with holiday-making urban people. So, too, is paying to camp or caravan in a farm field.

The second possible diversification route involves putting a farm to non-farming uses. Let us look at three possible ways.

■ Leisure and recreation – this might involve turning the farm into a shooting estate or modifying the fields for off-road driving and mountain biking.
■ Redevelopment – converting barns into housing, offices or industrial units, or using farmhouses for telecottaging in remote rural areas.
■ Energy – converting fields into wind or solar farms.

It is possible to mix a number of these different options to achieve a better income. One such mix might combine organic farming, running a farm shop and setting aside some fields for camping and energy generation. This sort of diversification helps to ensure that people will continue to live and work in rural areas.

SKILLS ▶ DECISION MAKING

ACTIVITY

In class, discuss the relative merits of the two types of farm diversification.

5.8 SUSTAINABLE RURAL LIVING

In this part of the topic you should recall and make use of what you have learned from your research into the changing use of rural environments.

First, we need to make clear what is meant by sustainable living. It has three related aspects.

- Environmental – a form of rural living that makes use of the environment and its resources, but does little damage to them. It makes sure that future generations will be able to continue to live there. The aim is to achieve a harmony between people and the land, i.e. the environment.
- Economic – what is needed are economic activities that support the local population and contribute to the regional or national economy. This might be by subsistence farming and self-sufficiency, or commercially by contributing products or services to the regional or national economy.
- Social – a form of rural living which offers a good quality of life for all, not just a select few. Important aspects are physical services such as clean water and sanitation, and social services such as education and healthcare. An unpolluted environment would be another significant aspect.

So, if we put all these aspects together, sustainable rural living needs a viable economy with a small carbon footprint and ecological footprint. In other words, a means of livelihood that does minimal damage to the environment and offers a good quality of life for all.

Let us look again at the two countries that figured in Part 5.6.

CASE STUDY: MAKING RURAL LIVING SUSTAINABLE IN CHINA

If its rural areas are to become sustainable, China needs to:
- increase the productivity of its agriculture in an environmentally-friendly way
- retain labour in the countryside to work on the farms
- raise the quality of rural life.

Extend the farmed area	Increase use of pesticides and herbicides
Enlarge farm holdings	Increase mechanisation
Improve irrigation	Improve plant and animal breeding
Set up crop rotation	Grow GM crops
Increase use of fertilisers	

▲ Table 5.3: Some possible ways of increasing food production

▲ Figure 5.25: The intensive use of Chinese rural space

Table 5.3 shows some of the more commonly used ways of raising the output of food. In China, all these actions are possible except one – extending the farmed area. Virtually all land capable of being farmed is already in use (Figure 5.25). The situation is not helped by the fact that large areas of agricultural land are being built over all the time.

There is much that could be done to improve the supply of home-grown food. For example, the government could encourage the amalgamation of farms into larger units. Mechanisation and modern farming techniques could be used to raise more food.

Droughts are a frequent hazard affecting Chinese agriculture. The South-North Water Transfer Project (see page 26) was supposed to increase the availability of irrigation water. But the supply of water to industry and homes comes first. This, and the fact that there is little intervention in farming, suggests that the Chinese government is focused on the secondary sector of its economy.

Adding to the poverty of rural areas is the fact there are few, if any, sources of income outside farming. In some parts of the world, for example Kenya, tourism is an important source of rural income. Is it possible that this might happen in China? At the moment, international tourism centres on cities; Beijing, Shanghai and Hong Kong, and the country's cultural heritage (the Great Wall, the Terracotta Army at Xi'an) attract tourists rather than its countryside and wildlife.

But, as China becomes more dependent on imported food, perhaps the government will increase efforts to help farming in its rural areas. Much needs to be done to improve the basics of rural living. This means providing better housing, a safe water supply, proper sanitation and pollution control. Only when the quality of life is raised, and rural areas become sustainable, will there be any chance of reducing rural-urban migration.

ACTIVITY

In groups, put together a list of the symptoms of rural poverty.

CHECK YOUR UNDERSTANDING

Summarise the challenges facing rural China.

SKILLS ⟩ INTERPRETATION

ACTIVITY

Which of the ways in Table 5.3 do you think would work best in a developed country?

CASE STUDY: MAKING RURAL LIVING SUSTAINABLE IN KENYA

In terms of improving agriculture, a significant move has already been made. Horticulture (the growing of vegetables, fruit and flowers) in Kenya has become one of the country's most important export industries. It accounts for two-thirds of total agricultural exports. Half a million people depend on the activity and the 135 000 workers who grow, cut and package fruit, vegetables and flowers for export. Kenya now supplies more flowers to the world market than any other developing country, except Colombia.

Seventy-five per cent of the flowers are grown on large farms, mainly around Lake Naivasha, and transported by air to markets in Europe. Soils around the lake are fertile and the lake provides water for irrigation. With intermittent droughts over the past 10 years, pressure on the lake has increased. The population of the area has risen from 50 000 to over 350 000 in 25 years, but services have not kept pace with this, resulting in untreated sewage being discharged into the lake. Fertilisers and pesticides also run off and pollute the lake. Water levels have

▲ Figure 5.26: Much of Kenya's horticulture is carried on under polythene.

dropped and local fisherman have seen a decrease in catches. With flowers very much a luxury, can their cultivation and their environmental costs really be justified (Figure 5.26)?

Much of the flower growing, as well as the production of fruit and vegetables, occurs on land previously used for subsistence farming. Large agribusinesses based in developed countries have bought the land. The promise of regular work and wages in the horticultural industry has easily persuaded many subsistence farmers to sell up. However, those farmers are beginning to see a big downside. With more land being used to grow crops for export, there is less land to grow food for the horticultural workers themselves and for the growing urban population. This is leading to food shortages. Food shortages mean higher food prices.

Another problem with Kenyan commercial horticulture is the food miles involved in transporting commodities to the major markets in Europe. Think of the huge carbon footprint of the aircraft used to fly fresh food and flowers to supermarkets in high income countries.

Although there is a downside to growing vegetables, fruit and flowers for the European market, some argue that it should not be stopped. To do this would deprive

Kenya of the foreign currency it needs to finance its economic development.

The rise of commercial agriculture does not solve the problem of feeding either the growing urban population or the people who live in the arid and semi-arid lands of Kenya. Poverty and food insecurity are acute. The physical environment is also highly degraded. Most of what little vegetation remains is being burned as fuelwood. The limited natural water supplies are being over-used.

Irrigation schemes are an obvious but expensive solution. So, too, is the provision of basic services, such as education, health, housing and safe water. These services can do much to improve the quality of rural life. A basic obstacle is a lack of money.

Where is the much-needed money to come from? Most likely from foreign aid. But corruption is making donor countries nervous. Another possibility is that the agribusinesses might be required to pay a share of their

▶ **Figure 5.27: Arid and semi-arid areas in Kenya**

SKILLS ▶ CRITICAL THINKING

ACTIVITY

Make a two-column table and list the costs and benefits to Kenya of commercial agriculture.

SKILLS ▶ INTERPRETATION

ACTIVITY

Write short notes about the pattern shown in Figure 5.27.

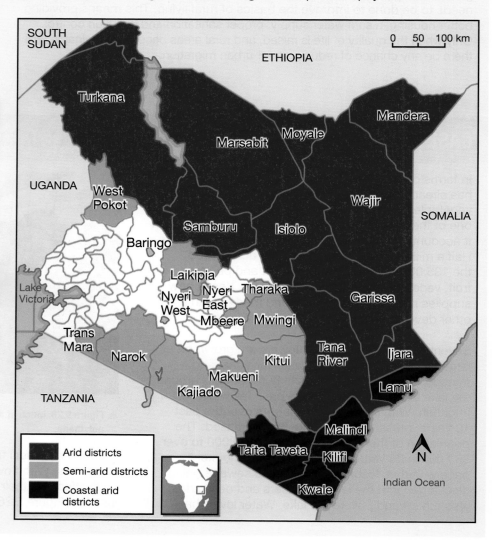

SOUTH SUDAN

ETHIOPIA

0   50   100 km

Turkana

Mandera

Moyale

Marsabit

UGANDA

West Pokot

Wajir

SOMALIA

Samburu

Isiolo

Baringo

Laikipia

Lake Victoria

Nyeri West

Nyeri East

Tharaka

Garissa

Mbeere

Mwingi

Trans Mara

Narok

Kitui

Tana River

Ijara

Makueni

Lamu

Kajiado

TANZANIA

Malindi

Taita Taveta

Kilifi

N

Kwale

Indian Ocean

Arid districts

Semi-arid districts

Coastal arid districts

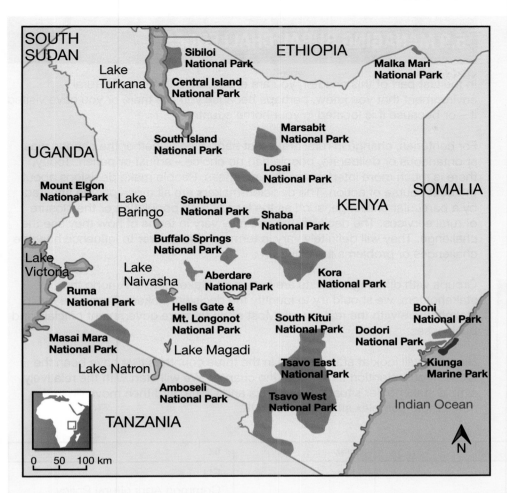

◀ Figure 5.28: Distribution of national parks and reserves

SOUTH SUDAN

ETHIOPIA

Lake Turkana

Sibiloi National Park

Malka Mari National Park

Central Island National Park

South Island National Park

Marsabit National Park

UGANDA

Losai National Park

SOMALIA

Mount Elgon National Park

Lake Baringo

Samburu National Park

KENYA

Shaba National Park

Buffalo Springs National Park

Lake Victoria

Lake Naivasha

Aberdare National Park

Kora National Park

Ruma National Park

Boni National Park

Hells Gate & Mt. Longonot National Park

South Kitui National Park

Dodori National Park

Masai Mara National Park

Lake Magadi

Lake Natron

Tsavo East National Park

Kiunga Marine Park

Amboseli National Park

Tsavo West National Park

Indian Ocean

TANZANIA

0   50   100 km

N

profits directly to the poorest rural areas. What is also needed is technical knowledge. Developed countries may be more willing to help here, particularly with improving soils and raising crop yields.

Rural Kenya does have one valuable natural resource – its wildlife. This has been exploited for many years by tourism. The tourist lodges of Kenya have a worldwide reputation. Providing for foreign tourists can be a source of income and employment. There are two limitations. Tourism is only really possible in the protected areas of national parks and game reserves (Figure 5.28). Much of the tourist industry is owned by foreign companies. So, much of the profits leave Kenya. Kenya needs to develop an eco-tourism model which emphasises the 'local' inputs.

As in China, much needs to be done to improve the quality of rural life. Attention needs to be focused on upgrading housing, installing piped water and sewage systems, and providing schools and healthcare.

The conclusion to be drawn from both case studies is that much still needs to be done. The rural areas of

China and Kenya are still far from being economically and environmentally sustainable. They lag behind urban areas in so many ways. The chances of reducing rural-urban migration, a major factor in rural poverty, remain very slim.

ACTIVITY

Find out more about one of the national parks shown in Figure 5.28.

SKILLS ▶ CRITICAL THINKING

ACTIVITY

Discuss, in class, how the challenges facing Kenya's rural areas differ from those of China.

5.9 MANAGING RURAL CHALLENGES

In the last part of this chapter, you are encouraged to think of a rural environment that you know, perhaps because you live there or you have visited it – or because it is located in your home country.

For centuries, change in rural areas just happened. Whether the change was spontaneous or deliberate, people had no choice – adjust or perish. Today, there is much more interference in rural affairs. People make decisions about the best course of action. The decision-makers are all those people affected by a particular challenge, such as the loss of rural population or the closure of rural services. The decision-makers may vary in terms of how they see the challenge. They will definitely vary in terms of their power to influence how the challenges or problems are solved.

Groups with differing interests are known as stakeholders. Among the stakeholders, we should try to identify the decision-makers. These are the stakeholders with the real power. Most often they are government officials and businesses.

Here, we will look at stakeholders in the three countries that have been the focus of our attention throughout the chapter. We will start with the relatively simple stakeholder situations in Kenya and China, and then move on to the much more complex situation in the UK.

SKILLS CRITICAL THINKING

ACTIVITY

Why are there no international stakeholders in rural China?

▼ Table 5.4: A selection of rural stakeholders in three countries

LEVEL	KENYA	CHINA	UK
International	IGOs World Bank OECD WHO NGOs Aid agencies Conservation organisations Tour operators Agribusinesses	None	EU Common Agricultural Policy Environmental directives
National	Government departments	Government departments	Government departments, especially DEFRA Voluntary organisations: National Trust English Heritage RSPB NFU CLA

Local	Farmers	Local communes
		Farmers
			Local government:
			  planning controls
			  services
			Local communities:
			  farmers
			  landowners
			  tourist boards

▲ Figure 5.29 A WHO immunisation programme in India

Let us take a brief look at some of the stakeholders in Table 5.4 and their particular interests.

INTERGOVERNMENTAL ORGANISATIONS (IGOS)

These involve representatives from member states. Some organisations with a particular interest in poor rural areas are listed here.

- The World Bank, which provides financial and technical help to countries to fight poverty and promote development.
- The United Nations Educational, Scientific and Cultural Organisation (UNESCO). One of its priorities is that every child should have access to education.
- The World Health Organization (WHO), which is primarily concerned with promoting healthcare and fighting disease (Figure 5.29).

NON-GOVERNMENTAL ORGANISATIONS (NGOS)

These are non-profit-making, voluntary organisations that operate at a local, national and international level. They are independent of government and often run as charities. Some have a special interest in rural areas and are listed below.

- Conservation organisations such as WWF (protection of wildlife and its habits) and Friends of the Earth (protection of the environment) (Figure 5.30).
- Aid agencies such as Oxfam, WaterAid and Save the Children (helping to provide basic services such as clean water, sanitation, schools and medical centres).

ACTIVITY

Name one more international conservation organisation and one more development aid agency.

▶ Figure 5.30: Friends of the Earth protesting about global warming

ACTIVITY

Can you name a national development aid agency in your home country?

GOVERNMENT DEPARTMENTS

Government responsibilities focus on the solution of rural problems, which include development and conservation issues. The way government works varies from country to country. Somewhere in the government structure, there will be a department or ministry with particular responsibility for what goes on in rural areas. In the UK, it is the Department for Environment, Food and Rural Affairs. In China, responsibility is divided between two ministries: Agriculture and Land and Resources. In Kenya, three departments share the responsibility: Agriculture; Livestock; Commerce and Tourism.

▶ Figure 5.31: A local community group protesting about the clearance of rainforest in Congo DR

LOCAL COMMUNITIES

Clearly, these are the people most affected by a change. The problem is that communities are made up of people and people have different viewpoints – some will support development and what they see as progress; others will favour conservation and fight change.

More stakeholders could be added to Table 5.4 on page 152. However, even in its incomplete state, the table suggests some possible conclusions.

Kenya: The influence of international organisations may be as strong, if not stronger, than that of the government departments. There are two main reasons:

■ many of the poorest rural areas get various sorts of outside help, much of it in the form of technical assistance
■ tourism and commercial farming are the two mainstays of the Kenyan economy; both involve marketing to international customers.

China: This country is governed by a communist regime. In theory, that should mean that power and decision-making rest with the people at a local level. However, this is not the case. Most aspects of Chinese life are controlled by a highly centralised national government.

Clearly, there are many more stakeholders here at national and local levels. They create a complex web of different views and interests. While all these stakeholders should have their say, it does mean that decision-making can be a long and drawn-out process.

SKILLS ▸ CRITICAL THINKING

ACTIVITY

For a rural area that you know, identify the main groups with an interest in its future.

CHAPTER QUESTIONS

END OF CHAPTER CHECKOUT

INTEGRATED SKILLS
During your work on the content of this chapter, you are expected to have practised in class the following:

- using world maps to show biomes
- using flow diagrams to represent the effects of different human activities on natural environments
- using photographs, marketing and social media to investigate diversification
- using socio-economic data to provide evidence that the quality of life has improved over time.

SHORT RESPONSE

1 State **two** factors that influence the global distribution of biomes. [2]

2 Identify **one** factor influencing the quality of rural life. [1]

3 State **two** examples of farm diversification. [2]

4 Identify **one** IGO that is particularly interested in rural areas. [1]

LONGER RESPONSE

7th 5 Explain **two** environmental impacts of deforestation. [4]

10th 6 Suggest reasons why the effects of natural hazards are often worse in rural areas. [4]

10th 7 Evaluate possible ways of halting the decline in remote rural areas. [8]

10th 8 For a rural area that you know, assess the parts played by different stakeholders. [8]

EXAM-STYLE PRACTICE

1 **a)** Identify the meaning of the term 'biome'. [1]

    A A local ecosystem
    B A reserve for indigenous tribes
    C A nature reserve
    D A large-scale community of plants and animals

    b) Name **one** input of an ecosystem. [1]

2 Identify the service provided by the tropical rainforest. [1]

    A Hardwood timber
    B Medicinal plants
    C Carbon sink
    D Minerals

3 Study Figure 5.1 (on page 130). Identify the main areas of tropical rainforest. [2]

4 **a)** State **one** factor attracting people to migrate from the countryside into urban areas. [1]

    b) Suggest **two** reasons for counter-urbanisation. [2]

5 Study Figure 5.22 (on page 145). Suggest why subsistence farming here is a struggle. [3]

7th 6 For a named country, suggest **two** possible ways of improving the sustainability of rural areas. [6]

7th 7 Evaluate the costs and benefits of changing from subsistence to commercial agriculture. [8]

[Exam-style practice total 25 marks]

6 URBAN ENVIRONMENTS

LEARNING OBJECTIVES

By the end of this chapter, you should know:

- The character of urbanisation and how this varies from place to place and over time

- The factors affecting the rate of urbanisation and the emergence of megacities

- The range of problems created by rapid urbanisation

- The factors affecting the urban land-use pattern

- The urban challenges facing a developed country

- The urban challenges facing an emerging or developing country

- Recent changes taking place in the rural-urban fringe

- Strategies for making urban living more sustainable

- The different groups involved in managing the challenges facing urban areas

Over half the world's population now lives in urban areas. Worldwide, urbanisation is changing where and how people live and work. The nature and rate of the process varies from place to place and over time. For many people, urbanisation brings benefits, such as a secure job and access to a range of services. But there are also serious costs such as congestion, discrimination, poor housing and environmental pollution. The nature of these costs varies nationally and internationally, often according to the level of development. For this reason, different strategies are needed to deal with the challenges of urban living. The hope is that urban living can be made more sustainable than it is today. But will this result in an improved quality of life for poorer urban residents?

6.1 URBANISATION AND ITS PROCESSES

An increasing percentage of a country's population living in urban settlements leads to growth in cities and towns and is called urbanisation. Urban settlements (towns and cities) differ from rural ones (hamlets and villages) in terms of:

■ their economies – residents make a living from manufacturing and services rather than agriculture
■ their size – they are larger in population and extent
■ the density of people and buildings, which is generally high
■ their way of life.

▶ **Figure 6.1: A modern cityscape at Doha, the capital of Qatar**

SKILLS   REASONING

ACTIVITY

Suggest reasons why there is so much high-rise development shown in Figure 6.1.

Figure 6.2 shows how the level of urbanisation (the percentage of the population living in urban settlements) varies across the globe. In general terms, it is the emerging and developed countries that show the highest levels of urbanisation. The lowest levels are found in Africa and South-East Asia.

▶ **Figure 6.2: Global distribution of urbanisation (2011)**

SKILLS   INTERPRETATION (NUMERICAL)

ACTIVITY

Make four general observations about the distribution shown in Figure 6.2.

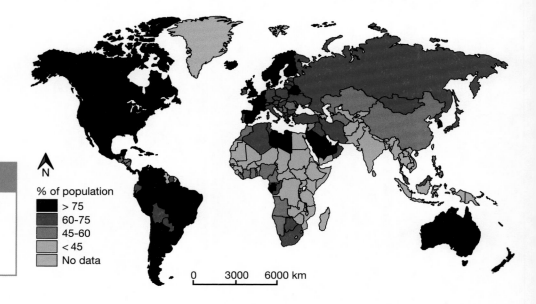

% of population
> 75
60-75
45-60
< 45
No data

0   3000   6000 km

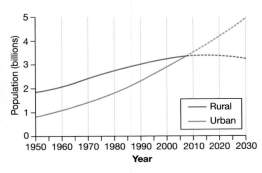

▲ **Figure 6.3: The changing global balance of rural and urban populations (1950-2030)**

Towns and cities are growing in number and size all over the world. While the world's population is increasing fast, the urban population is increasing even faster. Figure 6.3 shows that the world population more than doubled between 1950 and 2015, but that the urban population more than trebled. Half the world's population now lives in urban areas.

Present-day rates of urbanisation show the difference in the speed of growth between urban areas in developed countries and those in developing countries. The rate of urbanisation is much higher in the developing world (Figure 6.4). Present trends are expected to continue. However, the overall level of urbanisation remains higher in the developed world.

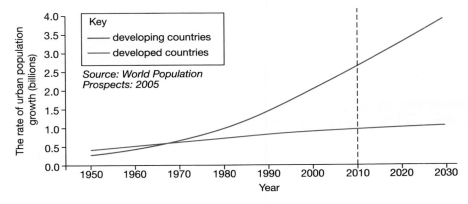

▶ **Figure 6.4: Urbanisation in the developed and developing worlds**

High rates of urbanisation occur in developing and emerging countries because:

- most new economic development in these countries is concentrated in the big cities
- **push-pull factors** are leading to high rates of rural-to-urban migration
- cities are experiencing high rates of natural increase in population.

In developed countries, rates of urbanisation are much slower because a large proportion of the population already lives in towns and cities. But the built-up areas of towns and cities continue to grow. Because of modern transport and communication, the urban way of life is gradually spreading into rural areas. In fact, the countryside and its settlements are experiencing what is referred to as rural dilution.

SKILLS　　**CRITICAL THINKING**

ACTIVITY

Can you suggest what some of the signs of rural dilution might be?

▶ **Figure 6.5: The urbanisation pathway**

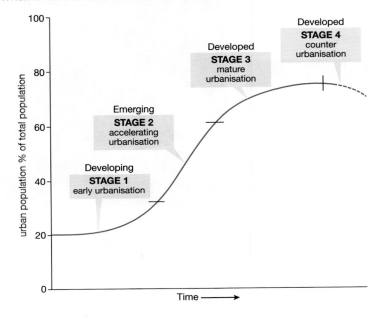

These differences between developed and developing countries encourage us to think of an urbanisation pathway. The pathway shows how the level of urbanisation changes over time (Figure 6.5). Countries become more urban as they develop economically. They gradually move from being a developing country towards becoming a developed country. In between, they pass through an 'emerging' stage. A country starts at a very low level of urbanisation (Stage 1). During the next stage the rates of economic development and urbanisation speed up (Stage 2). Later, as the pace of economic development slows, so too does the rate of urbanisation (Stage 3). The level of urbanisation begins to flatten out, often when roughly two-thirds of a developed country's population is living in urban settlements. Figure 6.5 shows that the pathway takes the form of an S-shaped curve. The curve may flatten as counter-urbanisation gains in strength.

Urbanisation is a process of change that converts rural areas and regions into urban ones. That change also involves other processes that affect the built-up areas of towns and cities as they grow.

PROCESSES

Urban settlements first appear as a result of **agglomeration**. This describes the concentration of people and economic activities at favourable locations, such as river crossing points, estuary mouths or close to a mineral resource (coal, iron or oil). In early times, defence was an important consideration when choosing location. As towns grow, they expand outwards by a process known as suburbanisation. This adds to the built-up area, but the building densities are generally lower than in the older parts of the town (Figure 6.6). The creation of these new suburbs, which are made up of houses, places of employment and services, is encouraged by:

■ improvements in transport that allow people to move easily between the new suburbs and the town centre
■ overcrowding, congestion and rising land prices in the older parts of the town
■ a general decline in the quality of the residential environment near the centre
■ the arrival of more people (mainly from rural areas) and new businesses.

▶ Figure 6.6: Suburban sprawl, on the edge of Paris

ACTIVITY

What do you think it is that appeals to those people who choose to live in the suburbs?

As a result of these two processes – agglomeration and suburbanisation – some towns grow into cities. Towns and cities located close to each other sometimes join together into one vast continuous built-up area known as a conurbation. As we shall see in Part 6.2, this scaling up in the size of urban settlements does not end there.

As urban settlements continue to prosper and grow, a new process sets in. People move out of the town or city altogether and live instead in smaller, often mainly rural settlements. These are called dormitory settlements because many of the new residents only sleep there. They continue to have links with the town or city they have left. They **commute** to the same place of work and still make use of urban services, such as shops, colleges and hospitals.

As cities and conurbations grow even bigger, a rather different process sets in. Rather than just moving out to suburbs and dormitory settlements, people and businesses move further out – either to smaller towns and cities or to rural areas. This process is known as counter-urbanisation.

▼ **Figure 6.7: Urban processes timeline**

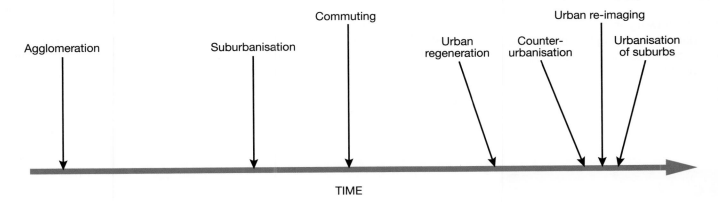

Having looked at a number of different processes, we can now put them into a time sequence (Figure 6.7). This sequence can be looked at in combination with the urbanisation pathway (Figure 6.5 on page 159). In the early stages of the pathway, urbanisation mainly involves agglomeration and suburbanisation.

In the later stages, although both agglomeration and suburbanisation continue, decentralisation and counter-urbanisation become more important.

Two new processes have begun to appear in the cities of the developed world. They are:

- **Urban regeneration** – this involves re-using areas in the old parts of cities abandoned as people and businesses have moved to the suburbs or beyond. This process allows **urban re-imaging** or **urban re-branding**, not just of city centres, but of whole cities (see Part 6.8). A well-known example of urban re-imaging is the regeneration of the deserted docklands and warehouses of London into upmarket offices and residential apartments. Former industrial settlements in the Ruhr (Germany) have also undergone 'facelifts'.
- Urbanisation of suburbs – these are typically areas of low density development. Today, however, as rural space is being eroded by urbanisation, governments are keen that more use is made of suburban areas. Vacant building plots and open spaces are being developed and large detached houses are replaced by flats and maisonettes. Suburbs are no longer protected as just residential areas. Shops and other services are located in the suburbs. So, suburban densities are now being raised to an urban level.

DID YOU KNOW?

Other well-known urban regeneration schemes in the UK include:

- Salford Quays, Manchester
- Cardiff Bay, South Wales
- Bradford, Yorkshire.

6.2 URBANISATION AND THE RISE OF MEGACITIES

RATE OF URBANISATION

From looking at Part 6.1, we can say that the main factors affecting the rate of urbanisation are as shown below.

SKILLS EXECUTIVE FUNCTION

ACTIVITY

Try drawing an annotated diagram showing the relationship between economic growth and population growth.

- The pace of economic development – it is economic growth that drives urbanisation. When the growth of the secondary and tertiary sectors is fast, so is the pace of urbanisation.
- The rate of population growth – economic growth needs an increasing supply of labour. The demand for more workers can be met in two ways: by either natural increase in the urban population or by rural-urban migration. The latter is usually by far the more important source of labour. It involves people being attracted by urban job opportunities and services, and by the perception that cities offer a better lifestyle.

There is a sort of 'multiplier effect' here with economic growth encouraging population growth. Population growth, in turn, makes more labour available, and more people need more services. The effect of this is more economic growth.

MEGACITIES

Fifty years ago, half of the world's 'ten largest cities' were located in developed countries and half in developing countries. By the year 2000, only two of the 10 were in the developed world.

▶ **Figure 6.8: Growth of millionaire cities (1900-2020)**

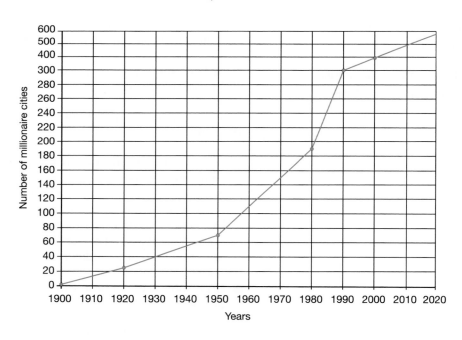

The global distribution of large cities has changed dramatically, but so has the total number of people living in them. The scaling up of city size is a feature of world urbanisation today. For many years, the millionaire city (a city of more than one million people) was considered a big city. In 1900 there were only two – London and Paris. Now there are about 500 (Figure 6.8). Today, the number of five-million cities has increased from eight in 1950 to over 50 (Figure 6.9).

▶ **Figure 6.9: The spread of cities with populations exceeding five million**

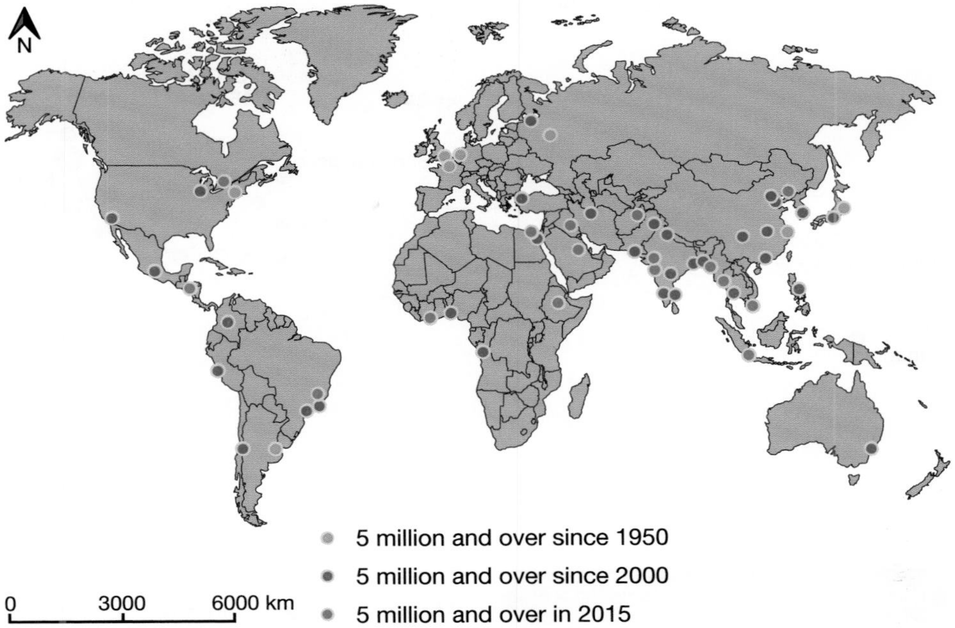

ACTIVITY

Study Figure 6.9. Analyse the changing distribution of five-million cities.

N

- 5 million and over since 1950
- 5 million and over since 2000
- 5 million and over in 2015

0   3000   6000 km

▼ **Figure 6.10: Factors encouraging the growth of megacities**

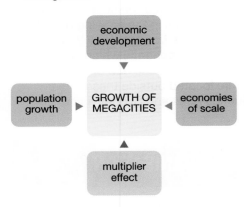

Can you think of any other downside to the growth of megacities?

More recently, the term **megacity** has been used to describe cities with populations of over 10 million. In 1970, there were just four of these, but there are now 35. Over half are in Asia.

What are the reasons for the growth of these megacities? Figure 6.10 shows four main factors. We have just looked at two of them – economic development and population growth. As regards population growth, and rural-urban migration in particular, young people are drawn to live in these megacities by the 'buzz' of feeling close to 'where it is all happening'. There is kudos and 'street cred' if living and working in such 'cool' places.

The other two factors encouraging megacities are listed here.

■ Economies of scale – there are advantages from cramming as much as possible into one megacity rather than into a number of smaller cities. Since distances within a megacity are less than between smaller cities, there are financial savings (economies of scale) in terms of transport. Communication between people and businesses (another economy) will be easier.

■ The multiplier effect – once a large city is prospering, it gathers a momentum which will carry it forward. This leads to yet more prosperity and growth. There are more jobs, so more people move into the city, which means there are more people who need goods and services. This creates more jobs and so the cycle goes on.

Megacities have a powerful attraction for both people and businesses. However, there is a downside. All the problems described in Part 6.3 (see next page) are present in megacities and are often even more acute. Their worst aspect is probably their impact at a national level. Megacities grow and prosper at the expense of towns, cities and regions elsewhere within the country. Megacities become powerful **cores** that create large **peripheries** around them.

▼ Figure 6.11: The network of global cities

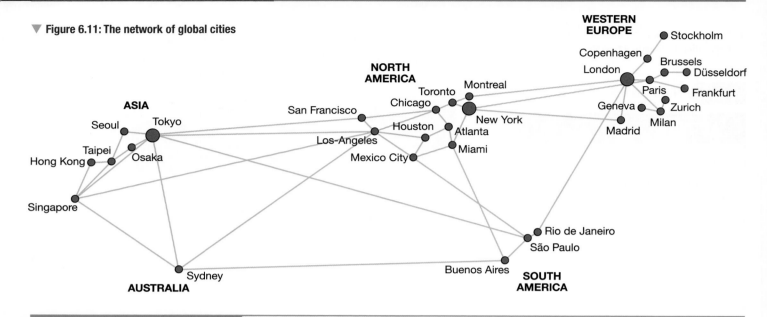

GLOBAL OR WORLD CITIES

CHECK YOUR UNDERSTANDING

What is the difference, if any, between a megacity and a world city?

Megacities are urban areas with populations greater than 10 million. Global or **world cities**, on the other hand, can be of any size. At present, there are 31 such cities (Figure 6.11). They all have populations over 1 million and seven of them are megacities (Buenos Aires, Hong Kong, Mexico City, New York, Rio de Janeiro, São Paulo, Seoul). What distinguishes a global city from a megacity? Global cities are recognised worldwide as places of great prestige, status, power and influence. All global cities are critical hubs in the growing global economy. The three most important global cities are London, New York and Tokyo. They are the financial centres of the global economy, which is why they have power and influence. Each of these three cities is the hub of a network of smaller global cities. Four global cities are located outside these three networks – Rio de Janeiro, São Paulo, Buenos Aires and Sydney – and are all in the southern hemisphere.

6.3 THE PROBLEMS OF RAPID URBANISATION

The world is rapidly becoming urbanised, and the pace of urbanisation is greatest in the emerging countries. For example, the population of the city of São Paulo in Brazil nearly doubled between 1970 and 2015. With a population of over 12 million, it is now the largest city in South America. Here, as elsewhere in the emerging world, rapid and often unplanned growth is creating a range of problems, mainly because of the speed of urbanisation.

CHECK YOUR UNDERSTANDING

Check that you understand the main causes of rapid urbanisation.

- Housing – much of the rapid growth of emerging cities is caused by people moving in from rural areas or smaller towns. When they arrive, there is nowhere for them to live, especially as many are looking for low-cost housing. Millions live in what were meant to be temporary **shanty towns** or **squatter settlements** (for more information, see Part 6.6). Even for people with money, the demand for housing exceeds supply. As a result, housing is expensive relative to wages. In general, because of poor transport, the most sought-after housing is close to the city centre with its shops and places of work.

- Access to water and electricity – often the provision of basic services does not keep up with the growth of population. As a consequence, not all parts of a built-up area will be provided with running water, sanitation or electricity. Many people rely on fires for cooking and lighting, and on polluted streams for water and sewage disposal.

- Traffic congestion and transport – the provision of proper roads and public transport is another aspect of city life that lags behind the growth in population. As a result, transport systems become overloaded and overcrowded, and traffic congestion is a major problem for everyone – rich or poor (Figure 6.12). The high numbers of vehicles also cause high levels of atmospheric pollution in cities. Many suffer regularly from smog (a mixture of smoke and fog).

- Health – often there are not enough doctors, clinics or hospitals to deal with the rapid increase in population. When large parts of a mushrooming city have little or no access to clean water or sanitation, diseases and infections (such as typhoid and cholera) spread quickly. Atmospheric pollution leads to widespread breathing problems.

- Education – rapid population growth also means a lack of schools. Although most cities manage to provide some primary education, not all children go on to secondary school. This is because of the cost and because many children have to work to help support their family.

- Employment – although people are attracted to cities for work, many are unable to find proper paid work. Instead, they are either unemployed or become part of the massive informal sector, surviving as best as they can. This includes selling goods on the street (Figure 6.13), working as a cleaner or shoe-shiner, cooking and selling food from home or by the roadside (see Part 4.5). Even where there is paid work in new factories, these are often many kilometres away from the shanty areas where most newcomers live.

▲ Figure 6.12: Traffic congestion in Mexico City

▲ Figure 6.13: Informal street-side worker in Dhaka, Bangladesh

▶ Figure 6.14: The growth of crime in Indian cities (1953-2011)

ACTIVITY

Write a short analysis of Figure 6.14.

ACTIVITY

Trying ranking the seven urban problems listed here, with the worst problem at the top. Give reasons to support your ranking.

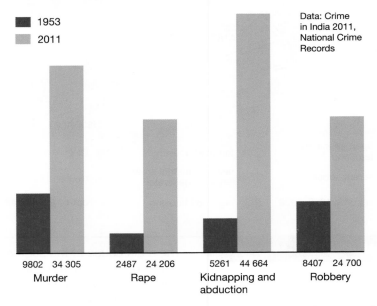

■ 1953
■ 2011

Data: Crime in India 2011, National Crime Records

| 9802 | 34 305 | 2487 | 24 206 | 5261 | 44 664 | 8407 | 24 700 |
| Murder | | Rape | | Kidnapping and abduction | | Robbery | |

- Social problems – given the crowded and unpleasant environment in which so many city dwellers live, it is not surprising that they also suffer from high crime rates. Murder, rape and robbery are three common crimes (Figure 6.14). The poorest areas are often inhabited by violent street gangs involved in drug trafficking.

- Environmental issues – there have already been several references in this chapter to environmental pollution. Traffic, industry and housing are among the worst polluters of air and water. But there is also noise (from road traffic, canned music in public places) and visual pollution (from unsightly advertising, graffiti). Cities produce large quantities of waste and waste disposal is another cause of environmental pollution. Also, a spreading urban area causes environmental damage of the surrounding countryside.

SKILLS ▷ INTERPRETATION

ACTIVITY

Study Figure 6.15. In what ways has the River Tyne influenced the urban pattern?

▼ Figure 6.15: The urban pattern of Newcastle-upon-Tyne, England

6.4 THE URBAN LAND-USE PATTERN

In this part of the topic you should recall and make use of what you have learned from your research into the changing use of urban environments.

Look at the built-up area of most towns and cities and you will see the same recurring features. There will be a central business district, industrial areas, a variety of residential districts, shopping centres, and so on. Figure 6.15 shows the segregation that is typical of cities in developed countries. What causes this segregation of different urban land uses? Why are the different land uses not jumbled up together?

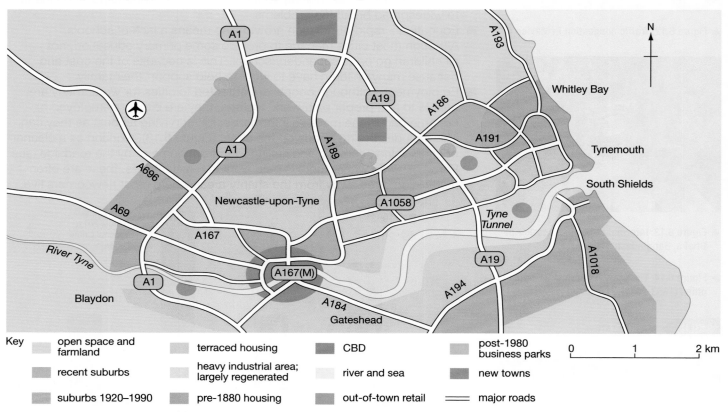

Key

open space and farmland	terraced housing	CBD	post-1980 business parks
recent suburbs	heavy industrial area; largely regenerated	river and sea	new towns
suburbs 1920–1990	pre-1880 housing	out-of-town retail	══ major roads

LAND VALUES

SKILLS ▷ INTERPRETATION

ACTIVITY

Describe the pattern of 'relief' created by urban land values, as shown in Figure 6.16.

The main cause of the segregation is the urban land market. As with the selling of any item, a particular site within the built-up area will normally be sold to the highest bidder. The highest bidder will be that activity that can make best use of a site. Usually retail shops can make the best financial use of land and property. To understand this, two related points need to be made clear: **land values** and locational needs.

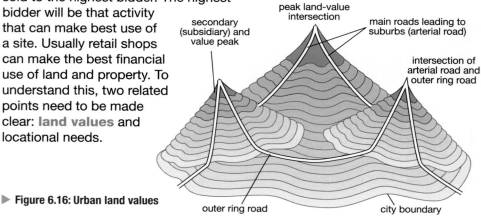

▶ Figure 6.16: Urban land values

First, land values vary within the urban area. Generally, they decline outwards from the centre, from the peak land-value intersection (Figure 6.16). However, relatively high land values are also found along major roads leading from the centre and around ring roads. Small land-value peaks occur where radial and ring roads cross each other. Businesses will pay extra for sites in these locations, because they are locations enjoying good accessibility.

LOCATIONAL NEEDS AND ACCESSIBILITY

Secondly, similar activities or land uses come together because:

- they have the same locational needs; these may be large amounts of space or being accessible to customers and employees
- they can afford the same general level of land values.

CHECK YOUR UNDERSTANDING

In which part of an industrial city in the developed world are most factories located?

Retailing and other commercial businesses (particularly offices) will cluster in and around the centre. This is the most accessible part of the built-up area. As a result of the clustering, they help define a central business district (CBD). Manufacturing also needs accessible locations for the assembly of raw materials and the dispatch of finished goods. However, it is a less capital-intensive use of space than shops or offices. Therefore, it has less buying power. So, manufacturing is found outside the CBD and most often along major roads that provide good accessibility and transport links. Housing is even less competitive on the urban land market. For this reason, housing is pushed further away from the centre. As land becomes cheaper towards the urban fringe, so houses become more spacious.

Because towns and cities grow outwards from a historic nucleus, they show concentric zoning – that is, a series of rings wrapping around the historic nucleus or core. So, in all towns in all cities no matter where they are located in the world, we see the same four features (Figure 6.17):

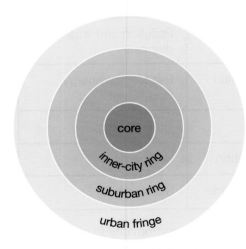

- a core – the oldest part of the city which normally contains the central business district and some of the earliest buildings
- an inner-city ring – early suburbs, so this has old housing and often some non-residential land uses
- a suburban ring – present suburbs with housing as the dominant land use
- an urban fringe – countryside being 'eroded' by the outward spread of the built-up area to provide space for housing and some non-residential uses.

We may make three more generalisations about the structure of cities as we move outwards from the core:

▲ **Figure 6.17: The four zones of a city**

- the general age of the built-up area decreases
- the style of architecture and urban design changes
- the overall density of development decreases.

CHECK YOUR UNDERSTANDING

What is the purpose of a model?

This urban model (known as the concentric zone model) with its four zones applies to virtually all towns and cities. What varies in different parts of the world is the character of each zone. That is, what goes on in them in terms of land use and the type of people living there.

RESIDENTIAL PATTERN

SKILLS   REASONING

ACTIVITY

Besides wealth, what else underlies the spatial sorting of people in residential areas?

People, like land uses, also become sorted within the urban area. They become segregated into groups within residential areas on the basis of their social class, type of occupation and ethnicity. People prefer to live close to those who they think are of the same status. However, the reason for most of these differences is personal wealth. The wealthiest people are able to buy smart and large homes in the best locations. The poorest people have to live in cramped or sub-standard housing in the worst residential areas. Many are unable to buy a home. Instead, they have to rent. Due to their limited means, they are forced to occupy only a small amount of space, and therefore live at high densities. This sorting within the residential parts of a city is an important aspect of the urban pattern.

6.5 URBAN CHALLENGES IN THE DEVELOPED WORLD

Many cities in developed countries face challenges which threaten both their general prosperity and the quality of urban living. Those challenges fall under three broad headings (Table 6.1).

▶ Table 6.1: Some challenges facing cities in the developed world

ECONOMIC	SOCIAL	ENVIRONMENTAL
Deindustrialisation	Social services and housing	Ecological footprint
Globalisation	Poverty and deprivation	Pollution and waste disposal
Food supply	Ethnic segregation	Resources: energy, land, water
Transport and traffic	Quality of life	Green space
Energy supply	Ageing population	Hazard risk
Service provision	Terrorism and crime	Sustainability

Clearly, the mix of challenges shown in Table 6.1 varies from city to city as each city is unique. Let us look at one of the world's leading cities – Hong Kong.

CASE STUDY: HONG KONG

Hong Kong has a land area of just over 1000 square km and a population of over 7 million. Although it has been a part of China since 1997, it is allowed to operate as an autonomous territory (Figure 6.18). This means that it has some political independence. Given its status as one of the world's top financial and business centres, it is not just a global city but a 'developed country' within an 'emerging' one.

ECONOMIC

Undoubtedly Hong Kong has benefited from globalisation. With its economic development based on trade and financial services, manufacturing has never been a major part of the economy. For this reason, it has not been much affected by de-industrialisation. Hong Kong is the world's eleventh largest trading 'country'. It is the world's largest re-export centre.

▶ **Figure 6.18: Map of Hong Kong**

ACTIVITY

Describe the site of Hong Kong.

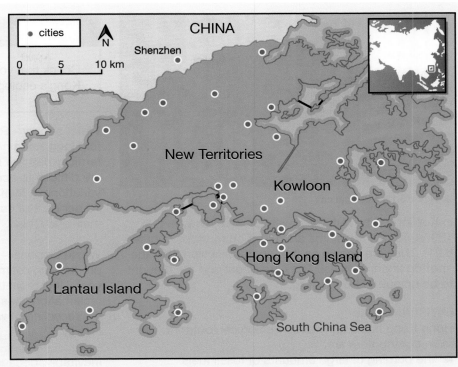

These re-exports are products made outside the territory, especially in mainland China, which are distributed via Hong Kong. Hong Kong has few natural resources.

FOOD
The territory of Hong Kong has little arable land, so agriculture is a relatively unimportant part of the economy and contributes only 0.1 per cent of its GDP. Inevitably, it has to import most of its food. For reasons explained in Part 5.8 (page 148), most of that food comes from outside China.

ENERGY
All cities need a cheap and reliable supply of energy. The energy supply in Hong Kong is mainly in the form of electricity. Around 75 per cent of this is generated by the burning of fossil fuels (coal much more than natural gas). Nearly a quarter of the electricity is generated by nuclear power, but this is imported from mainland China.

TRANSPORT
In so many cities, traffic congestion is a major issue. Fortunately, Hong Kong has a well-developed transport network and movement around this densely populated territory is relatively easy. Over 90 percent of the 11 million journeys made every day are on public transport. This is a remarkably high percentage. Public transport is provided by an integrated network of railways, buses and ferries.

SOCIAL
Hong Kong provides good social services, particularly education and healthcare.

HOUSING
As one of the world's most densely packed places, the only way Hong Kong can grow is upwards. It has become a 'vertical city' (Figure 6.19).

About half of Hong Kong's residents live in public housing, which is rented from the government. This housing is mainly in very high-rise apartment blocks. With regard to private housing, Hong Kong is ranked as the third most expensive city in the world. The effect of this is to reduce social mobility. It is extremely difficult for people to move from the public to the private sector. This also underlines another feature of Hong Kong's population – its social polarisation. Extreme wealth, as well as poverty and deprivation, are found in the compact urban area.

Ethnic segregation is a factor contributing to social polarisation. About 94 per cent of Hong Kong's population is of Chinese descent. The remaining 6 per cent is made up mainly of people from the Philippines, Indonesia, Nepal and India. Most of them are employed by the wealthy as domestic servants. They are concentrated in the Central and Western districts, as well as Kowloon City, Yau Tsim Mong and Yuen Long.

There are also Britons, Americans, Canadians, Japanese and Koreans living in Hong Kong. They work in the city's commercial and financial sector and live in the wealthier districts.

POVERTY AND DEPRIVATION
Despite its overall prosperity, Hong Kong contains pockets of poverty and deprivation. The most obvious symptoms are the slums. There used to be many slums

▲ Figure 6.19: A vertical city

in Kowloon City, but these were cleared away in the 1990s. Today, the slums are found on rooftops (Figure 6.21). They are known as the 'penthouse slums' or the 'rooftop shanty towns'. They are illegal and are thought to be unique to Hong Kong.

ENVIRONMENTAL

Hong Kong has a fairly deep ecological footprint. The main contributors are:

- the burning of large amounts of fossil fuels
- the need to reclaim large amounts of land from the sea to create new space for urban growth (Figure 6.20); this has damaged the marine environment
- the smog that drifts across the territory from industrial developments on the other side of the Pearl River delta
- the disposal of waste (see right).

▲ Figure 6.20: Reclaiming land in Hong Kong for further expansion

ACTIVITY

Check that you understand what makes Hong Kong's ecological footprint so deep.

Water supply – Providing an adequate water supply for Hong Kong has always been difficult because the territory has few rivers and lakes. Groundwater sources are difficult to access. A large population and seasonal variations in rainfall mean that Hong Kong has to import 70 per cent of its water supply from Dongjiang River in mainland China.

Waste disposal – Hong Kong's 11 million residents produce an estimated 6.4 million tonnes of waste a year. Until recently, this was dumped in landfill and land reclamation sites. Attempts are being made to increase the recycling of waste, as well as converting waste into energy.

Inevitably, because of its size and economic success, Hong Kong is a huge and concentrated consumer of resources. These resources range from land and water to energy and recreational space.

All this adds to the deepness of Hong Kong's ecological footprint. Hong Kong's challenge lies in becoming sustainable.

▲ Figure 6.21: Penthouse slums are built on the top of high-rise buildings.

6.6 URBAN CHALLENGES IN THE DEVELOPING AND EMERGING WORLDS

Many of the challenges listed in Table 6.1 also apply to cities in emerging and developing countries. In this section, we focus on four different types of challenge. To illustrate this, we will look at two countries and their capital cities: Mexico (representing emerging countries) and Kenya (representing developing countries).

SQUATTER SETTLEMENTS

SKILLS ▸ INTERPRETATION

ACTIVITY

How many cities in Table 6.2 are located in emerging countries?

▸ Table 6.2: Percentage living in squatter settlements and slums in selected cities

SKILLS ▸ REASONING

ACTIVITY

Explain the link between overcrowding and hazards.

DID YOU KNOW?

Squatter settlements are known by different names around the world:

■ favelas (Brazil)
■ barriadas (Latin America)
■ bidonville (North Africa)
■ bustees (Indian subcontinent).

The speed of urbanisation in emerging and some developing countries is fast and accelerating. Most people who migrate to cities come from poor rural areas in search of work. There are no houses for them, so they build homes on the only land available. This is usually in areas of no economic value, on the edge of town, along main roads or on steep slopes. These makeshift housing areas are known as shanty towns. In many instances, people build on land that they do not own, or build on land without permission to build. As a result, such areas are also known as squatter settlements. Another worldwide term for these are slums. Table 6.2 shows just how important these developments are in providing basic shelter for the urban poor. Remember, too, that some people are so poor that they live out in the open on the streets.

CITY	% OF CITY POPULATION
Nairobi, Kenya	est. 55
Mexico City	46
Lima, Peru	36
Caracas, Venezuela	35
Koolgata, India	33
Rio de Janeiro, Brazil	27
Jakarta, Indonesia	25
Santiago, Chile	25

Many of the areas where shanty towns are built are unsafe. They may be prone to flooding or landslides or are in heavily polluted locations. Other hazards include fire, crime and the spread of disease, which are often linked to the overcrowding.

Usually, shanty towns are not serviced with pipe water and waste disposal. The actual dwellings are made out of scrap materials such as packing boxes, metal and plastic sheeting. However, for many people, even living in a shanty town and working in the informal economy can be better and offer greater opportunities than the life they left behind in rural areas.

▸ Figure 6.22: A shanty town in Mexico City

THE INFORMAL ECONOMY

DID YOU KNOW?

For more on informal employment, see Part 4.5.

Squatter settlements offer the urban poor, particularly rural-urban migrants, some form of shelter. As explained in Part 4.5, the informal economy offers those people a means of survival. This is particularly the case in cities such as Mexico City and Nairobi, where there is both unemployment and underemployment. Here people face a simple but grim choice. Either do something that will earn you and your family just enough to survive on, or slowly die of starvation.

▼ Table 6.3: Informal employment as percentage of total non-agricultural employment in a sample of countries (2012)

COUNTRY	INFORMAL EMPLOYMENT AS % OF NON-AGRICULTURAL EMPLOYMENT
Brazil	42
China	33
India	84
Indonesia	72
Kenya	78
Mexico	54
Pakistan	78
Sri Lanka	62
Vietnam	68
Zambia	70

In Kenya, 78 per cent of all working people are engaged in the informal sector; the figure is even higher in Nairobi and other urban settlements. In Mexico, the percentage is half that much lower.

You will notice in Table 6.3 that the percentage importance of the informal sector in Kenya is much higher than in Mexico. It is likely that the percentages are even higher than the national averages in Nairobi and Mexico City.

▲ Figure 6.23: Informal activities in Nairobi

URBAN POLLUTION

It is not surprising that pollution is a problem in squatter settlements. They lack piped water, proper sanitation and waste disposal. The burning of fuelwood pollutes the air. There are also other sources of pollution outside the slums. Congested traffic causes air pollution. Manufacturers exploit lax controls and pollute both air and rivers with their discharges. Adding to the pollution of the urban environment are smells and noise. Visual pollution is caused by garbage in the streets, graffiti and unsightly buildings.

LOW QUALITY OF LIFE

▼ Table 6.4: Quality of Life Index (2015); the top country was Denmark = 201.53

COUNTRY	QUALITY OF LIFE INDEX
Mexico	142.85
Chile	140.86
India	106.28
Brazil	94.75
China	94.59
Kenya	87.48
Sri Lanka	32.35
Thailand	49.48
Sri Lanka	32.35
Vietnam	31.48

Deprivation is a term widely used when discussing a low quality of life. This is defined as a standard of living below that of the majority of people in a particular city, region or country. It involves all those hardships just described – poor and congested housing, a lack of secure employment and a polluted living environment. Deprivation also involves a lack of access to a range of other things, such as a proper diet, schooling, medical treatment, leisure and recreation.

The Quality of Life Index (QLI) takes into account eight different variables. These include safety, healthcare, cost of living and pollution. Table 6.4 shows national differences for selected countries. It reflects the difference between emerging and developing countries (compare Mexico and Kenya).

Two questions cannot be answered by reading this table.

■ How would city values compare with national values – higher or lower?
■ Would the difference between Mexico City and Nairobi be roughly the same as the difference between their national values?

SKILLS CRITICAL THINKING

ACTIVITY

Discuss these two questions in class.

6.7 DEVELOPMENTS ON THE URBAN FRINGE

In this part of the topic you should recall and make use of what you have learned from your research into the changing use of urban environments.

The areas where the green fields and open spaces of the countryside meet the built-up parts of the towns and cities is known as either the rural-urban fringe or the urban fringe. Here, countryside is being lost by the outward growth of towns and cities, particularly their suburbs. The **greenfield sites** of the open land around the edge of a city are in great demand for housing, industry, shopping, recreation and the needs of public utilities, such as reservoirs and sewerage works.

One reason for urban growth and change on the urban fringe is a feeling of dissatisfaction with the city. Here are some push factors.

- Housing is old, congested and relatively expensive.
- There are various forms of environmental pollution – air quality is poor, and noise levels are high.
- Companies find that there is a shortage of land for building new shops, offices and factories. So, any unused land is costly.

There are also pull factors on the urban fringe.

- Land is cheaper so houses are larger.
- Factories can be more spacious and have plenty of room for workers to park their cars.
- Closeness to main roads and motorways allows for quicker and easier customer contacts.
- New developments on the outskirts are favoured by the personal mobility allowed by car drivers.

What else is happening around the urban fringe? Besides the appearance of new housing estates, there are significant non-residential developments. We shall focus on four of these.

RETAIL PARKS

In developed countries, there has been a great increase in out-of-town retailing, with large purpose-built superstores and shopping centres located at or just beyond the urban fringe. The number of superstores has increased dramatically in the UK since 1980. It is easy to understand why. More people own their own cars. The large car parks are free. Access is easy because the shopping centres are located next to main roads and motorway junctions. In contrast, city centre shoppers face traffic congestion and expensive parking. The larger out-of-town centres have shopping malls which are bright and modern with everything under one roof. Other facilities (such as multi-screen cinemas or bowling alleys) are often within the shopping centre or are located close by, so there is something there for all the family.

Often, due to good main roads, the big retailing developments are serving customers drawn from more than one town or city.

INDUSTRIAL ESTATES

These are areas of modern light industries and service industries with a planned layout and purpose-built road networks.

SKILLS CRITICAL THINKING

ACTIVITY

Are you able to add to the lists of push and pull factors?

DID YOU KNOW?
For more information about changes in the urban fringe, see Part 4.2 (page 102).

BUSINESS PARKS

These are areas created by property developers to attract firms needing office accommodation, rather than industrial units. These often include leisure activities such as bowling alleys, ice rinks and cinemas.

SCIENCE PARKS

These are usually located close to a university or research centre with the aim of encouraging and developing high-tech industries and quaternary activities.

THE GREENFIELD VERSUS BROWNFIELD DEBATE

ACTIVITY

Think about your own set of values. Do you think that the countryside should be protected at all costs, or should more be released for urban growth?

Not everyone is happy with the continued loss of countryside around the towns and cities of developed countries. Many environmentalists believe that new developments should be built on brownfield sites – land within the built-up area that has been abandoned and is now lying idle – rather than on greenfield sites.

The question of where to build (on greenfield sites at the edge of the built-up area or on brownfield sites well inside the built-up area) arises in connection with a range of urban land uses – housing, retailing, industries and offices. With all land uses, there are arguments for and against each type of site. As Table 6.5 shows, each has its advantages and disadvantages.

SITE	ADVANTAGES	DISADVANTAGES
BROWNFIELD		
	Reduces the loss of countryside and land that might be put to agricultural or recreational use.	Often more expensive because old buildings have to be cleared and land made free of pollution.
	Helps to revive old and disused urban areas.	Often surrounded by rundown areas so does not appeal to more wealthy people as residential locations.
	Services already in place.	Higher levels of pollution; less healthy.
	Located near to main areas of employment.	May not have good access by road.
GREENFIELD		
	Relatively cheap and rates of house building are faster.	Valuable farmland, recreational space and attractive scenery lost.
	The layout is not hampered by previous development so can easily be made efficient and pleasant.	Development causes noise and light pollution in the surrounding countryside.
	Healthier environment.	Wildlife and their habitats lost.
	Proximity of countryside, leisure and recreation.	Encourages further suburban sprawl.

▲ Table 6.5: The advantages and disadvantages of brownfield and greenfield sites

There is no clear winner in this debate. It all depends on the particular land use. Housing is fairly flexible in terms of where it might be built, but shops, offices and industries need more specific locations.

It also depends on the circumstances of the town or city. Is the green space really valuable? Are there serious problems and high costs involved in reusing the brown space?

SKILLS ▶ CRITICAL THINKING

ACTIVITY

Should the development of brownfield sites be given priority over the development of greenfield sites? Give your reasons.

6.8 MAKING URBAN LIVING MORE SUSTAINABLE

In this part of the topic you should recall and make use of what you have learned from your research into the changing use of urban environments.

We have seen what sustainable living means in some rural environments (Part 5.8, page 148). Now we turn to the urban environment. What does sustainable living mean here, and how might it be achieved?

Making urban living more sustainable may be achieved through a range of different activities:

CHECK YOUR UNDERSTANDING

What is the link between sustainability and ecological footprints?

- using renewable rather than non-renewable resources
- using energy more efficiently
- relying on public rather than private transport
- improving the **physical infrastructure** – clean water and proper sanitation
- improving social services and access to them
- improving the quality of life, particularly of the urban poor.

Most of this adds up to a single challenge – to reduce the ecological footprints of towns and cities. Added to that is the aim of reducing social inequalities (Figure 6.24).

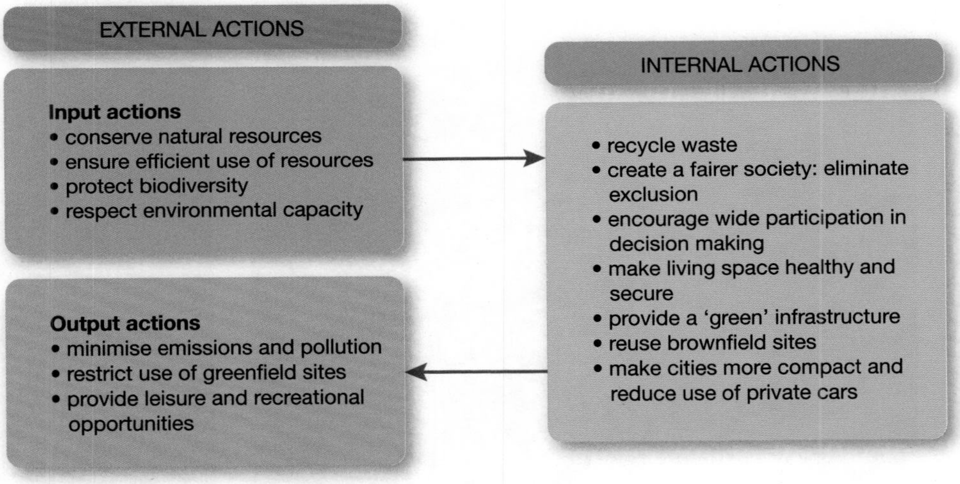

▶ **Figure 6.24: Making cities and urban living more sustainable**

EXTERNAL ACTIONS

INTERNAL ACTIONS

Input actions
- conserve natural resources
- ensure efficient use of resources
- protect biodiversity
- respect environmental capacity

- recycle waste
- create a fairer society: eliminate exclusion
- encourage wide participation in decision making
- make living space healthy and secure
- provide a 'green' infrastructure
- reuse brownfield sites
- make cities more compact and reduce use of private cars

Output actions
- minimise emissions and pollution
- restrict use of greenfield sites
- provide leisure and recreational opportunities

CHECK YOUR UNDERSTANDING

Check that you understand the differences between the three types of action shown in Figure 6.24.

In this section, we look at attempts in three different parts of the world to make urban living more sustainable. The hope is that others will be inspired to follow these examples. Each example shows a different aspect of urban sustainability:

- Masdar City, Abu Dhabi – energy
- Curitiba, Brazil – transport and greening
- Urban green partnership, Sri Lanka – poverty and greening.

CASE STUDY: MASDAR CITY, ABU DHABI (UAE)

Masdar City claims to be one of the world's most sustainable cities. Building of this eco-city started in 2008 (Figure 6.25). Now nearing completion, it will house 40 000 people, and 5000 people will commute there every day, either to work or study.

The sustainability of the project is based on reducing the consumption of energy and water, and reducing the production of waste. All of the energy supply is renewable. Nearly all comes from solar power. It is generated by rooftop solar panels and one of the largest photovoltaic plants in the Middle East. The design of the streets makes good use of the coolness provided by the prevailing winds. The buildings combine traditional Arabic architectural techniques with modern technology. The orientation of buildings and their design minimises the need for air conditioning, heating and artificial light (Figure 6.26). So, the city's carbon footprint is a small one.

Masdar City has installed smart water consumption systems in all the city's buildings. They are designed to consume 54 per cent less water than the United Arab Emirate's average building. In addition, 75 per cent of hot water is provided via thermal receptors fixed on top of the buildings. Water comes from desalinisation plants using renewable energy.

Waste is reduced as near as possible to zero, through encouraging changes in behaviour (stressing the need for recycling), and controlling the types of material that can be used within the city (maintaining a war on plastic and polythene).

Citizens are required to attend five hours of sustainability education each year. Much of this education aims to change lifestyles and make them more environmentally-friendly.

Masdar City has become a leader in research and education in sustainable and clean technologies. This has attracted an increasing number of businesses keen to market new clean technological discoveries.

The achievements of Masdar City are impressive. But there are two limitations:

- the large sums of capital needed to set it up
- it is only suitable for those parts of the world with access to plentiful supplies of renewable energy.

CHECK YOUR UNDERSTANDING

What is 'green' about Masdar City?

▲ Figure 6.25: An aerial view of Masdar City

▲ Figure 6.26: Some of Masdar's energy-efficient housing

CASE STUDY: CURITIBA, BRAZIL

Curitiba in south-west Brazil is now a city with a population of well over 2 million (Figure 6.27). It is widely recognised as a good example of urban planning. The planning started in 1968 when the city's population was already 430 000.

The Curitiba Master Plan was first aimed at improving transport. Five main roads converging on the city centre were converted into dual carriageways separated by a central two-lane carriageway for exclusive use by express buses (Figure 6.28). Triple-articulated buses provide fast, efficient and cheap transport and this has persuaded people to leave their cars at home. During the rush hours, buses run every 60 seconds and are always full. The network is now used by 70 per cent of the city's inhabitants. One fare allows passengers to travel anywhere on the network. Buses now use biofuels and this has reduced pollution.

But there is more to eco-friendly Curitiba than just its transport network.

The town is virtually surrounded by parks for public recreation. These parks also stop favelas (shanty towns) being established on the urban fringe, although they have not been completely successful in preventing the growth of favelas. Flooding is a regular hazard in Curitiba. However, lakes created within the parks are now providing an effective flood control service. There are also parks within the city. The grass here is controlled by grazing sheep.

The city recycles its waste and has done so since 1980. It has set up a pioneering waste disposal system. Waste is collected through a network of 'cambo verde' sites. Curitiba now recycles over two-thirds of its waste. People are paid for the garbage they collect not in money, but in fruit and vegetables. The scheme has been very effective in improving conditions in the slums. More recently, attention has turned to dealing with the slums. Guided by the city authority, a new self-help suburb is being built to replace demolished slum dwellings.

The 'greening' of Curitiba was the idea of Jamie Lerner, a former mayor. It is a model of sustainable urban planning from an emerging country. But its ideas have been taken up mainly in developed countries. The ultimate verdict on Curitiba is provided by a recent survey which found that 99 per cent of Curitiba's residents are happy with their city.

▼ Figure 6.27: Curitiba

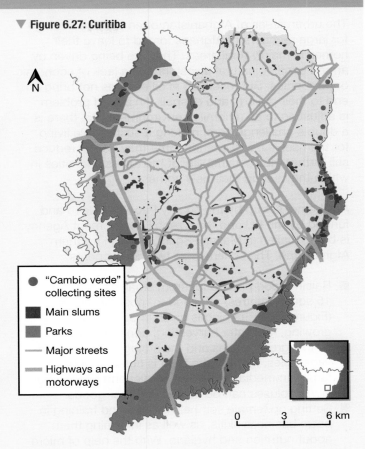

- ● "Cambio verde" collecting sites
- ■ Main slums
- ■ Parks
- — Major streets
- ▬ Highways and motorways

0　3　6 km

CHECK YOUR UNDERSTANDING

What actions have been involved in the 'greening' of Curitiba?

▲ Figure 6.28: In Curitiba, bus lanes are separated from cars.

ACTIVITY

Find Curitiba in your atlas. Name the nearest major city.

CASE STUDY: URBAN GARDENS, AFGHANISTAN

The urban areas of Afghanistan often provide a refuge for large numbers of Afghans forced to leave their homes in mainly rural areas. They are being driven by armed conflict, natural disasters and a lack of economic opportunities. But resettling in cities does not bring an end to their poverty and despair. An added problem is adjusting to an unfamiliar environment. So, there is a double challenge here: making a sustainable living for themselves, and ensuring that they do not reduce still further the sustainability of the towns and cities in which they set up their new homes.

A scheme, set up by the charity People in Need and funded by the EU and the Czech Development Agency, is being trialled in Mazar-i-Sharif, a city in northern Afghanistan. The scheme has two components.

- Raising food production from small plots (10 to 15 square metres) attached to most dwellings (Figure 6.29). Training is given in basic crop growing, rainwater harvesting, constructing simple greenhouses and the storage of harvested vegetables. Most of the food is consumed by the family (immediately benefiting diet and health), but any surpluses can be traded for other goods.
- Setting up female self-help groups and training in basic business skills, as well as informing them about nutrition and hygiene. With the help of micro-loans, new businesses are being set up. Examples are shops, spinning and making clothes, and rearing poultry. All this is helping to empower women in society.

The Mazard-i-Sharif project provides a possible model for other urban areas in Afghanistan. It is helping to remedy food insecurity, unemployment and poverty. It is making urban areas less reliant on food from rural areas. It is also making urban areas more self-sufficient and, as a result, a little more sustainable.

▲ Figure 6.29: Mazar-i-Sharif: a) an urban garden, and b) a dressmaking enterprise

SKILLS ▶ REASONING

ACTIVITY

Explain the significance of the training in nutrition and hygiene.

6.9 MANAGING URBAN CHALLENGES

The groups involved in managing any urban challenge will vary according to the specific challenge. The roles of those groups will also vary. So, it is dangerous to generalise. It would be better if we look at one challenge common to so many cities at all levels of development – slums. Certainly, they are a challenge in all three cities – Hong Kong, Mexico City and Nairobi.

▶ Figure 6.30: A woman sits in a bulldozed squatter camp, near Harare (Zimbabwe).

SKILLS DECISION MAKING

ACTIVITY

Which of the five slum management options would you support?

LOCAL

There are probably five different slum-management options:

■ 1 bulldoze and clear away (Zimbabwe)
■ 2 clear away but relocate (Brazil)
■ 3 redevelop (Brazil)
■ 4 improve by self-help or site-and-services schemes (Philippines)
■ 5 ignore (India, Bangladesh).

Inside the brackets are examples of countries that have taken action of this kind.

Let us assume that we are dealing with this challenge in a country with a democratic government. What interest groups or stakeholders are likely to be involved in choosing one of the five management options? As in Part 5.9, the groups of interested people and organisations may be divided into three. Let us start at the grassroots (local) level.

The stakeholders are as shown here.

■ Slum residents – they will obviously support anything that improves their conditions. But how great is their demand for change and improvement? How 'vocal' are they?
■ Residents living nearby – they will be keen on Options 1, 2 and 3, but do they speak 'with one voice' as a well-organised protest group?
■ Utility suppliers (water, waste disposal and electricity) – can they provide these services and be sure of receiving payments for them?
■ Representatives in parliament or on city councils – how keen are they that slum-dwellers should have improved living conditions? How keen are they to put pressure on city and national governments to do something? There may well be votes to gain if they support a particular option.
■ City councils – they are likely to have most influence in the choice of option. They are in charge of the resources that might be needed. But they can be lobbied and pressured by well-organised groups or stakeholders.
■ Landowners and property developers – are there sites away from the squatter settlements on which replacement high-density housing might be built? Would the landowners be keen to sell? Might they be persuaded to sell cheaply?
■ Employers – what might their reactions be if the squatters were relocated? Is it in their interests to keep the supply of cheap labour where it is?
■ Planners – what is their vision of the future? Might it be of a city without squatter settlements?

NATIONAL

The stakeholders at national level are as shown here.

- Government – are there financial resources to support a programme of action? Apart from Option 5, which is likely to be the cheapest option? Which option is likely to gain them most votes? Is slum improvement one of their policies?
- National charities – most countries have home-grown charities or small charities set up by foreigners to help people living in slums. Magic Bus is a good example of a charity set up by foreigners. The help is in the form of volunteers and practical help rather than money.

INTERNATIONAL

The stakeholders at international level are as shown here.

- International charities – there are many international charities supporting actions aimed at particular squatter settlement problems. Examples include Oxfam, CAFOD, Christian Aid and MSF. They are mainly to do with Option 4. Their help is targeted at: water and sanitation; health and infectious diseases; education; food; and child labour. They will need to assess the urgency and scale of the problem before taking any action.
- Inter-governmental organisations – IGOs like the World Bank, UNESCO and the World Health Organization which sponsor projects aimed at helping the poor. Support is usually channeled through national governments. Corruption can get in the way of delivering the help to where it is most needed.

▶ Figure 6.31: A favela in São Paulo, Brazil

The final point to make is this. The choice of management option will be largely determined by local stakeholders. National governments may either support or oppose that choice. International stakeholders may become involved once the decision has been taken. Charities are more likely to help if the management choice is either Option 3 or Option 4.

SKILLS ▶ REASONING

ACTIVITY

Illustrate how and why governments may have different views on slums.

CHAPTER QUESTIONS

END OF CHAPTER CHECKOUT

INTEGRATED SKILLS
During your work on the content of this chapter, you are expected to have practised in class the following:

■ using world maps to show trends in urbanisation over the last 50 years
■ interpreting photographs and maps to investigate the impacts of rapid urbanisation
■ using satellite images to identify different urban land uses
■ using and interpreting socio-economic data
■ using quantitative and qualitative information to judge the scale of variations in environmental quality.

SHORT RESPONSE

1 Identify **two** problems resulting from rapid urbanisation. [2]

2 State **two** social problems facing cities in the developed world. [2]

3 Identify the least-urbanised continent. [1]

4 State the name of the process aimed at bringing new life to old urban areas. [1]

LONGER RESPONSE

5 Explain what is meant by the term 'deprivation'. [4]

6 Explain why urban gardening is being encouraged in developing countries. [4]

7 Examine the reasons why people become segregated into groups in the residential areas of cities. [8]

8 Assess **two** ways of improving living conditions in squatter settlements. [8]

EXAM-STYLE PRACTICE

1 **a)** Identify the meaning of the term 'counter-urbanisation'. [1]
   A Population movement away from the largest cities
   B Population growth on the edge of urban areas
   C Population movement into inner-city areas
   D Population growth in the capital city

   b) Identify **one** feature that distinguishes urban from rural settlements. [1]

2 Identify the minimum population size of a megacity. [1]
   A 1 million
   B 5 million
   C 10 million
   D 15 million

3 Study Figure 6.2 (on page 158). Identify the year in which the global population was equally divided between urban and rural. [1]

4 **a)** State **one** factor affecting the urban land-use pattern. [1]

   b) Explain **two** reasons for the development of squatter settlements. [2]

5 Suggest **three** ways of making urban living more sustainable. [6]

6 Examine the reasons why people choose to live in suburbs. [4]

7 Evaluate the arguments in favour of using brownfield rather than greenfield sites. [8]

[Exam-style practice total 25 marks]

7 FRAGILE ENVIRONMENTS AND CLIMATE CHANGE

LEARNING OBJECTIVES

By the end of this chapter, you should know:

- The distributions and characteristics of fragile environments
- The causes of desertification and deforestation
- The causes of climate change
- The impacts of desertification
- The impacts of deforestation
- The negative effects of climate change
- How technology can help to reduce the threat of desertification
- The different approaches to the management of the tropical rainforest in a named region
- Different responses to global warming and climate change from individuals, organisations and governments

This chapter is about processes and pressures that are making some environments increasingly 'fragile'. Desertification of the semi-arid areas of the world is partly the outcome of population growth exceeding environmental carrying capacities. Deforestation of the tropical rainforest is the result of a large-scale and unsustainable exploitation of its resources. Global warming and climate change are increasing the fragility of some environments. What, if anything, can be done by way of new technologies and different management strategies to reduce the fragility? What are the responsibilities of individuals, organisations and governments?

7.1 FRAGILE ENVIRONMENTS

CHARACTERISTICS

The well-being of the Earth's physical environments is of vital importance to us all. Our living standards and our health depend on the quality of those environments. However, natural environments are very **fragile**. There is a delicate balance between non-living parts (climate, rocks, soils) and living parts (plants, animals). Natural hazards, such as fires, high winds and volcanic eruptions, have always disturbed environments and made them more fragile (Figure 7.1). However, in most cases, environments have recovered. For thousands of years, people have been making use of environmental resources to provide food, fuel and building materials. They have done so without causing too much environmental damage. Early people lived in harmony with the environment.

▶ Figure 7.1: Fire is a natural hazard, as well as a human environmental hazard.

DID YOU KNOW?

Many environments are 'fragile' because of the delicate nature of most ecosystems. Look at Part 5.2 (page 132) to help you to understand why.

However, the growth of the world's population today threatens to disturb the fragile balance of environments. People have disturbed 90 per cent of the Earth to some degree or another. It is hard to find areas of truly natural wilderness untouched by human activity.

CHECK YOUR UNDERSTANDING

In what ways do people threaten ecosystems?

DISTRIBUTIONS

The fragility of environments is closely related to the pressure that is put on them. The ecological footprint is a measure of the mark that humans make on the natural world. It considers how much land and sea are required to provide us with the water, energy and food we need to support our lifestyles. If the Earth's resources were shared equally, it is believed that a 'fair share' for everyone would be a little less than 2 hectares of the globe. The UK has an ecological footprint of about 5.5 global hectares per person. This means that if everyone in the world consumed resources at the rate of people in the UK, we would need two more planets to sustain the world's present population. Figure 7.2 shows how the ecological footprint varies around the world. It gives us an impression of where we might expect environments to be made fragile by people.

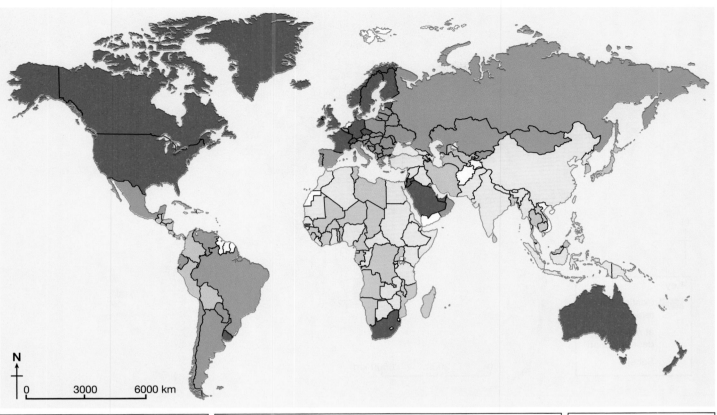

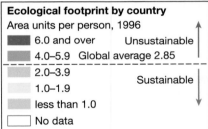

Ecological footprint by country
Area units per person, 1996

■	6.0 and over	Unsustainable
■	4.0–5.9	Global average 2.85
■	2.0–3.9	Sustainable
■	1.0–1.9	
■	less than 1.0	
□	No data	

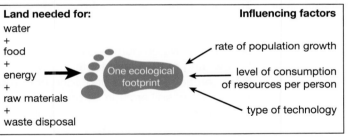

Land needed for:
water
+
food
+
energy
+
raw materials
+
waste disposal

Influencing factors

rate of population growth

level of consumption
of resources per person

type of technology

One ecological footprint

Source:

Living Planet Report

Reproduced with
permission from WWF

▲ **Figure 7.2: Global variations in the
ecological footprint**

The global picture given by Figure 7.2 is more complicated than this. It does
not take into account the fact that developed and emerging countries import
large amounts of food and energy. So, these imports deepen the ecological
footprints of the countries supplying them.

The global distribution of fragile environments is most influenced by the
impact of three processes – desertification, deforestation and climate
change. The three are linked because the first two are both the causes and
consequences of climate change. Together, they are making environments
more fragile. Let us now look at first two processes.

ACTIVITY

Identify the six countries on the
southern half of the map that
have unsustainable ecological
footprints.

DESERTIFICATION

Desertification is the term used to describe how once-productive land
gradually changes into a desert-like landscape. The process is not
necessarily irreversible and, as Figure 7.3 shows, it usually takes place in
semi-arid land on the edges of existing hot deserts. The worrying message
illustrated by Figure 7.3 is that large areas of the world are at risk from
desertification. The most conspicuous includes much of southern Asia,
the Middle East and North Africa.

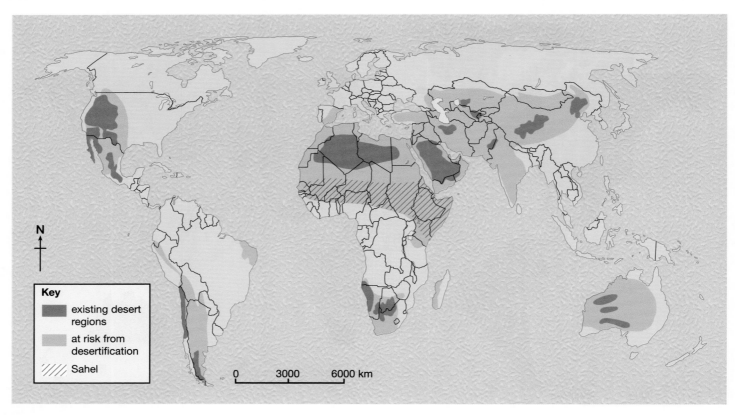

Key

■ existing desert regions

■ at risk from desertification

/// Sahel

0   3000   6000 km

▲ Figure 7.3: Areas at risk from desertification

The main characteristics of desertification (Figure 7.4) are:

- absence of surface water
- dried up watercourses and ponds
- lowering of the water table
- vegetation becomes degraded or completely lost
- increased soil erosion as bare soil is exposed to wind
- increase in salt content of the soil
- soil becomes less usable
- increasing presence of dry, loose sand.

SKILLS ▸ INTERPRETATION

ACTIVITY

How many of the symptoms of desertification are visible in Figure 7.4?

▶ Figure 7.4: Desertification in Burkina Faso

DEFORESTATION

Deforestation is the cutting down of trees. Many primary forests in temperate countries have almost disappeared after centuries of logging (cutting down trees for timber) and land clearance, usually to plant crops (Figure 7.5). This deforestation has been most severe in the deciduous forests of the warm temperate parts of Europe, China and the USA. As yet, the coniferous forests of the cold temperate regions of North America and Eurasia remain relatively untouched. However, the same cannot be said of the world's tropical rainforests. Here the speed of deforestation has alarmed scientists and conservationists. The future welfare of tropical rainforests is an important environmental issue.

▼ Figure 7.5: Global distributions of current forests and areas of forest loss

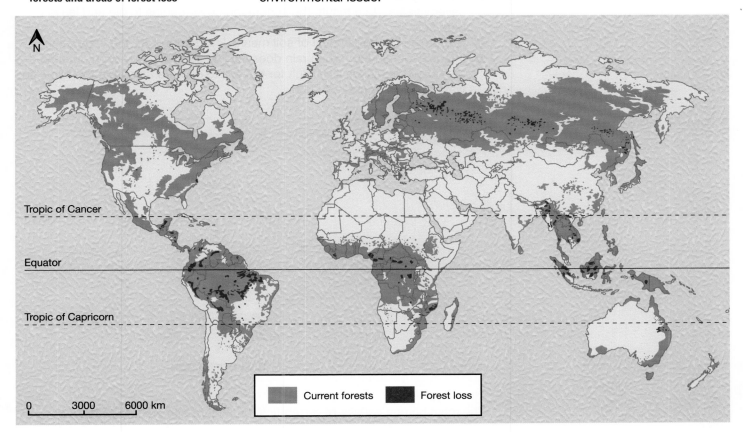

Current forests   Forest loss

The characteristics of deforestation are obvious (Figure 7.6). They depend, for example, on whether the forest has been subject to clear felling or selective felling. In other words, has the forest been completely cut down (clear felling); or have only the best trees been extracted so that some trees remain (selective felling)? The symptoms of deforestation will also depend on how long ago the deforestation took place. Freshly sawn tree trunks and burned saplings suggest recent action, while large fields used to grow crops surrounded by a fringe of surviving trees indicate that the deforestation took place decades or even centuries ago. As soon as land loses its tree cover, the soil is exposed to the erosive effects of wind and rain.

▲ Figure 7.6: A recently deforested landscape

SKILLS INTERPRETATION

CHECK YOUR UNDERSTANDING

Explain the link between deforestation and soil erosion.

ACTIVITY

Look at Figure 7.5. Briefly describe where the greatest losses of tropical forest have occurred.

7.2 CAUSES OF DESERTIFICATION AND DEFORESTATION

DESERTIFICATION

Desertification is the result of both natural and human causes. Natural causes (Figure 7.7) are shown below.

■ Changing rainfall patterns – rainfall has become less predictable over the past 50 years and the occasional drought year sometimes extends to several years. As a result, the vegetation cover begins to die and leaves bare soil.
■ Soil erosion – the removal of soil means less support for the vegetation.
■ Intensity of rainfall – when rain does fall it is often for very short, intense periods. This makes it difficult for the soil to capture and store the rain – so water resources are reduced.

▶ Figure 7.7: Natural causes of desertification

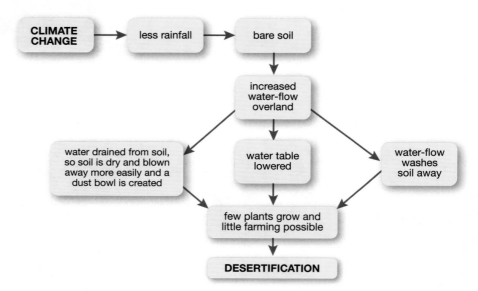

The main human causes of desertification (Figure 7.8) are shown below.

■ Population growth – rapid population increase puts more pressure on the land to grow more food.
■ Migration – as desertification takes hold in one area, local people migrate elsewhere in search of food and water. Unfortunately, wherever they settle, they increase the population pressure on the environment.
■ Overgrazing – too many goats, sheep and cattle can destroy vegetation. This is often most common around water holes in desert fringe areas.
■ Over-cultivation – intensive use of marginal land exhausts the soil and crops will not grow.
■ Deforestation – trees are cut down for fuel, fencing and housing. The roots no longer bind the soil, leading to soil erosion.

These problems are worse for people in sub-Saharan countries because of years of civil war. Crops and animals have been deliberately destroyed resulting in famine and widespread deaths.

▶ **Figure 7.8: Human causes of desertification**

SKILLS ▶ DECISION MAKING

ACTIVITY

In groups, discuss which of the five human causes of desertification is thought to be the most significant. Give your reasons.

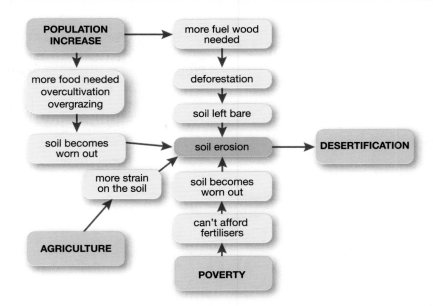

As with many other fragile environments, it is usually a combination of different factors which results in damage to a specific area.

It is estimated that about 20 per cent of the world's population has to cope with the effects of desertification in over 60 countries. One region most at risk is the Sahel region of Africa, an area south of the Sahara Desert stretching the width of the continent. It makes up a large part of sub-Saharan Africa, the poorest region of the world.

CASE STUDY: DESERTIFICATION OF THE SAHEL, AFRICA

The Sahel is a narrow belt of land in central Africa. It borders the southern edge of the Sahara Desert (Figure 7.9). The Sahel has a semi-arid climate as shown in Figure 7.10. Temperatures are always hot and there is a long dry season from June through to January. There is just enough rainfall for grasses to grow, as well as some shrubs and trees in this harsh environment.

The world biomes map (Figure 5.1 on page 130) shows that the Sahel region is classified as savanna. The natural vegetation of savanna is a mixture of grassland, trees and shrubs. However, the mix of each of these three types of vegetation changes as we move northwards from the tropical rainforest. The climate becomes drier, so that wooded savanna gives way to grassland with occasional shrubs. Eventually, on the margins of the hot desert, the savanna is nothing more than thin grassland. It is these areas on the margin of the present desert which are among those most at threat from desertification.

On the equatorial edges of savanna regions there are also more animals. As the trees thin out towards the middle of savanna regions, large herds of wild animals

like wildebeest, antelope and zebra are found. On the drier desert edges there is far less wildlife. Rainfall is seasonal and unpredictable and it is very dry for much of the year (Figure 7.10). Here you find nomadic herders who move from place to place with goats and cattle in search of water and grazing.

In the Sahel, some years have less rain than others. Fewer grasses grow and trees die. The landscape becomes much more like desert (Figure 7.11).

Climate change is one cause of desertification. Desertification is also speeded up by human activity. Until the 1960s, water was plentiful and crops and livestock did well in the Sahel. Then there was an increase in population. More and more trees were felled to provide fuel and building materials. It was possibly this was responsible for the climate becoming drier.

As this was happening, people tried to grow the same crops and rear the same number of animals. This quickly led to over-cultivation and overgrazing and much ground was laid bare. The absence of vegetation meant that no humus was added to the soil. Without

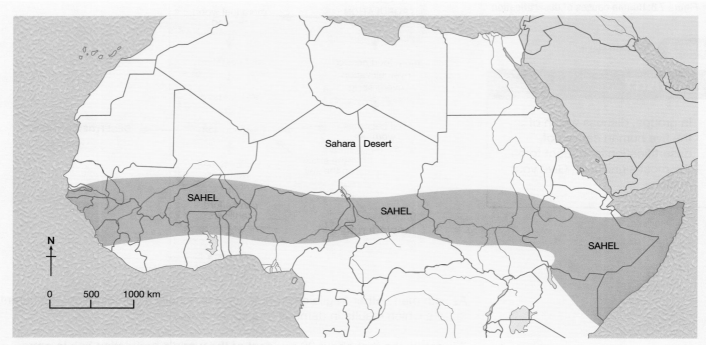

▲ Figure 7.9: Location and present extent of the Sahel

SKILLS INTERPRETATION

ACTIVITY

Using an atlas, identify all the countries shown in Figure 7.9 that are part of the Sahel.

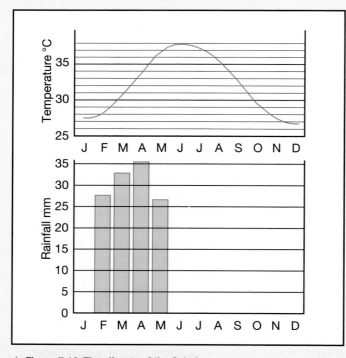

▲ Figure 7.10: The climate of the Sahel

humus, the soil was able to hold little or no water. As the soil dried out, it was quickly eroded by wind and occasional flash floods.

Since the 1970s, crop failures have become almost an annual event. Over 100 000 people have died of starvation. Even more people have migrated to less arid areas. Millions of animals have died.

A variety of techniques might prevent the further spread of desertification and also rehabilitate the land already damaged. One successful method of catching rain when it falls is a simple technique set up by Oxfam in Burkina Faso (see Figure 7.33 on page 205). Small stone walls are built following the slope of the land, which then act as dams when the rain falls; this stops surface water run off and allows water to sink into the soil. This simple, inexpensive method can increase crop yields by up to 50 per cent.

Most scientists think that people are the root cause of desertification. However, recent research using satellite images is showing that some areas suffering from desertification are now showing signs of recovery. They are beginning to receive more rainfall. Could it be that natural changes in climate over short time periods are the root cause of desertification, and not people? Nobody can be sure at present. One thing is certain. The semi-arid lands are fragile environments and people must use them with great care.

SKILLS ▸ DECISION MAKING

ACTIVITY

Do you think that migration is a possible solution to the problems of the Sahel? Justify your answer.

▶ Figure 7.11: Searching for water in the Sahel

DEFORESTATION

Why are rainforests being cut down?

Commercial logging/timber extraction (globally 26%)
Only valuable trees are chopped down (selective logging) but as they fall, they damage other trees. Even more damage is caused by 'clear-felling', where other trees are also chopped down and chipped for pulp.

Agriculture (globally 32%)
Areas of tropical rainforest have been cleared for plantations growing a single crop such as rubber or coffee. Plants and grassland are grown, which huge herds of cattle graze on for a few years before another area is cleared for seeding with grass.

Road building
Roads have been built through rain forests to enable minerals, timber, cattle and crops to be moved easily. Roads also bring in new settlers who clear areas for farming.

Mining
Large areas of forest are cleared for the open-cast mining of minerals such as iron, gold and copper.

Land for farmers
Land allows farmers to grow their own food and the wood that is cut down provides them with fuel. It stops overcrowding in other parts of the country.

HEP (hydroelectric power)
Rivers are dammed and huge areas of forest are flooded as a result.

▶ Figure 7.12: The reasons why rainforests are being cut down

Areas of tropical rainforest are cleared for a variety of reasons (Figure 7.12). Trees are felled for timber and sometimes for their medicinal drugs. Large areas have been deforested to make land for farming, housing and industry. Mining and hydroelectric power (HEP) schemes have also led to land clearance. Sometimes, the logging and land clearance is illegal or without any sort of control. Some developing country governments have encouraged the clearance of forests because:

- the revenue earned from selling timber, minerals and the sources of medicinal drugs helps to pay off debts and to fund economic development
- more land is needed to house and feed the fast-growing populations in countries such as Brazil and Malaysia.

DID YOU KNOW?

The causes of deforestation in Brazil were examined in Part 5.3 (page 134). It is important that you now re-read that section.

7.3 THE CAUSES OF CLIMATE CHANGE

The Earth's climate has changed many times during its history. For example, the most recent Ice Age ended only 12 000 years ago. Many people therefore believe that climate change must be due to natural causes. They argue that variations in the amount of solar radiation received by the Earth are the main cause. These variations result from:

■ eccentricity in the shape of the Earth's orbit around the Sun
■ changes in the tilt of the Earth's axis (obliquity)
■ wobbles as the Earth spins on its axis (precession).

As a result of increasing and decreasing amounts of solar energy reaching the Earth, the Earth heats up and cools down. This leads to global climate changes.

▼ **Figure 7.13: Milankovitch's solar cycles**

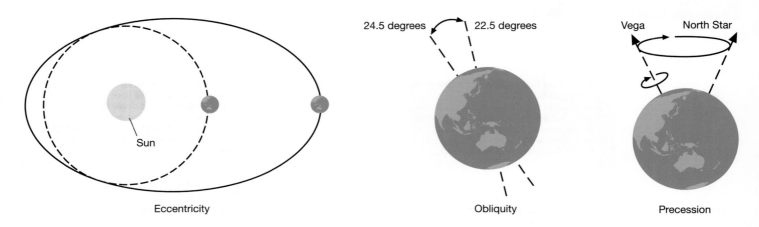

Eccentricity

Obliquity

Precession

Milankovitch's theory (Figure 7.13) is that every 100 000 years, these three changes combine in a particular way. The net effect is to significantly reduce the amount of solar radiation. The Earth experiences a glacial period.

There are possibly two other natural causes of climate change.

■ Volcanic activity – large-scale volcanic eruptions eject huge amounts of dust and ash into the atmosphere. Such mega eruptions can block out solar radiation. If this happens, the result is a lowering of global temperatures to the point that there is a glacial period.
■ Cosmic material – if huge meteors and asteroids reach the Earth's surface, they, too, can eject large qualities of dust into the atmosphere. The impact is the same as with volcanoes. One theory argues that the extinction of the dinosaurs at the end of the Jurassic period was the result of a huge comet suddenly changing global climates.

Only in more recent times have we made and kept accurate measurements of our weather and climate. This has allowed scientists to identify changes and trends based on recorded evidence at varying levels of detail over the past 100–150 years.

▶ **Figure 7.14: Global temperature changes since 1850**

ACTIVITY

What does Figure 7.14 tell you about changes in global temperatures between 1850 and 2000?

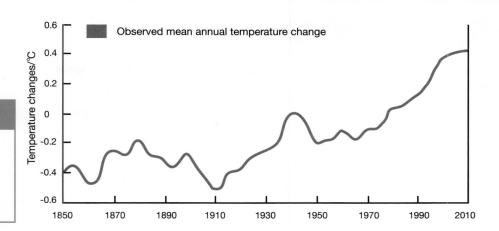

Figure 7.14 shows how temperatures have changed – and increased – since records began. On average, global land temperatures are 1°C higher now than they were at the end of the 19th century. Measurements over the past 100 years have shown an average rise of 0.7°C. The 1990s saw some of the highest temperatures ever recorded. A number of leading organisations have predicted a rise of up to 4.5°C by the end of the 21st century. This may not seem to be very much. But a recent study of temperatures in the Permian period, 250 million years ago, suggests that a 6°C rise in temperature led to the extinction of 95 per cent of the species living at that time.

It is now widely believed that the present rise in global temperatures is largely the result of human activity. This is mainly due to the release of greenhouse gases, mostly via human activity. Table 7.1 lists the main greenhouse gases and how they are released into the atmosphere.

▼ **Table 7.1: Greenhouse gases and their sources**

GREENHOUSE GAS	SOURCE
Carbon dioxide (CO_2)	Released when fossil fuels (coal, oil, etc.) are burned, for example in electric power stations and by motor vehicles. Burning of wood as a fuel releases CO_2. Deforestation – trees remove CO_2 from the atmosphere. So, if trees are removed, CO_2 levels rise.
Methane (CH_4)	Decay of organic matter – waste in landfill sites, animal manure, large areas of crops.
Nitrous oxides (N_2O)	Burning of fossil fuels and use of artificial fertilisers.
Chlorofluorocarbons (CFCs)	Gases released via aerosols and coolants in fridges, freezers and air conditioning systems; also in some types of packaging and insulation.

Check that you know the four greenhouse gases and their sources. Which is the most common?

It is perhaps the similarity between the rise in global temperatures and in global carbon dioxide emissions that has convinced so many that there is a causal link between carbon dioxide emissions and global warming; carbon dioxide gas is thought to be a major contributor to the **greenhouse effect**.

THE GREENHOUSE EFFECT

▶ **Figure 7.15: Annual greenhouse gas emissions by sector (2014)**

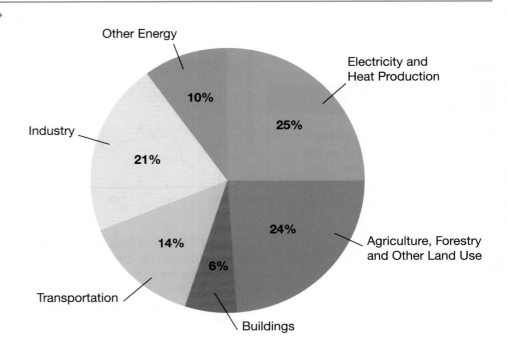

Greenhouse gases occur naturally in the atmosphere. However, they can also be produced and released by a range of activities (Figure 7.15). Carbon dioxide is the main cause for concern as it accounts for nearly three-quarters of all greenhouse gas emissions. Factory emissions, burning **fossil fuels** like oil, coal and gas in power stations, and exhaust emissions from motor vehicles are the major sources of greenhouse gas emissions. The main producers are developed countries, with the USA alone responsible for 36 per cent of all greenhouse gas emissions. However, the effects are felt by everyone.

In a greenhouse, the Sun shines through the glass and warms up the plants inside. When the Sun stops shining, the heat does not disperse; it is trapped inside the greenhouse. In the same way, heat is trapped inside the Earth's atmosphere. During the day, radiation from the Sun heats the Earth (Figure 7.17). At night, clouds often trap this heat as it radiates back out. Gases in the atmosphere also trap this heat. This is the greenhouse effect. In recent years, the amount of greenhouse gases has greatly increased. They build up in the atmosphere, preventing heat from radiating back out. This is thought by many scientists to be the main cause of the gradual increase in world temperatures known as global warming.

It is important to note, however, that some scientists are not convinced that people and their activities are the cause of today's global warming. They argue that recent changes in global temperatures are simply part of natural climate change. The Earth's climate has changed a great deal over geological time. Even in historic times, there have been quite sudden changes in temperature – alternations of hot and cold periods.

Finally, the three processes examined so far in this chapter are not the only ways in which natural environments are being upset and made more fragile. But all practices that upset environments have to do with exploiting the Earth. Pollution of the air, land and water is a major culprit (Figure 7.16). So, too, are various forms of river and coastal management. These practices are often carried out with good intention, but they still disturb the workings of sensitive natural systems.

▶ **Figure 7.16: Oil spills are a particularly damaging form of human hazard.**

ACTIVITY

Give examples of how coastal management can upset coastal ecosystems.

▶ **Figure 7.17: The greenhouse effect**

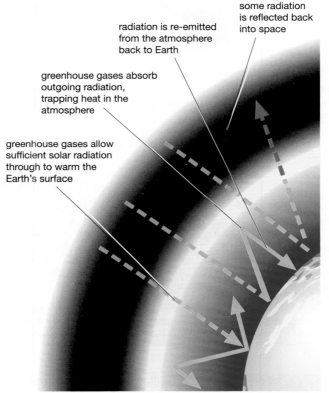

some radiation is reflected back into space

radiation is re-emitted from the atmosphere back to Earth

greenhouse gases absorb outgoing radiation, trapping heat in the atmosphere

greenhouse gases allow sufficient solar radiation through to warm the Earth's surface

▲ **Figure 7.18: The end product of soil erosion, Namibia**

7.4 THE IMPACTS OF DESERTIFICATION

Once desertification has started, five things follow more or less in sequence.

- Soil – as the plant cover gradually disappears, soils cease to be replenished by the nutrient cycle (see page 133). A decline in soil fertility continues as people try to grow crops and raise livestock. The exhausted soil dries out and soon is a victim of erosion. Eventually, it becomes dust and is dispersed by the wind (Figure 7.18).
- Reduced agricultural output – the decline in the condition of the soil, plus the increasing shortage of water, means that the agricultural productivity of the land is lowered. Crops fail and livestock die.
- Malnutrition, famine and starvation – the decline in agricultural output leads

ACTIVITY

Draw an annotated diagram showing the links between the five impacts of desertification.

to food shortages (famine) and a poor diet. People can survive famine for some time, but after a while the human body cannot cope, leading to starvation and possibly death. The young and the elderly are badly affected. Malnutrition, famine and starvation are all symptoms of overpopulation.

■ Migration – eventually the situation becomes so bad that people face a difficult choice – either die where you are or migrate in the hope of finding somewhere where you can raise enough food to survive.

■ Conflict – a problem with migration is that it often leads to conflict. Migrants arriving and settling in an area where other people are only just surviving will not be welcomed.

An added problem with these four consequences is they are linked in a downward chain reaction. The situation becomes progressively worse. That is, unless some drastic actions are taken (see Part 7.7).

7.5 THE IMPACTS OF DEFORESTATION

▼ Figure 7.19: Some problems caused by deforestation

These impacts have already been covered in Part 5.3 with respect to the Amazon rainforest in Brazil. It is important that you re-read that case study. The following is a more general discussion of the impacts of deforestation.

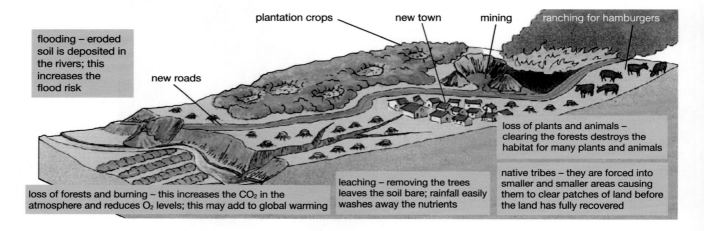

flooding – eroded soil is deposited in the rivers; this increases the flood risk

plantation crops   new town   mining   ranching for hamburgers

new roads

loss of plants and animals – clearing the forests destroys the habitat for many plants and animals

native tribes – they are forced into smaller and smaller areas causing them to clear patches of land before the land has fully recovered

loss of forests and burning – this increases the CO_2 in the atmosphere and reduces O_2 levels; this may add to global warming

leaching – removing the trees leaves the soil bare; rainfall easily washes away the nutrients

Check that you understand each of the five problems shown in Figure 7.19.

Figure 7.19 illustrates some of the physical consequences of deforestation. There are negative impacts, not just on vegetation, but also on drainage, soils and wildlife.

Today there is concern about the rate of deforestation. Figure 7.20 gives some information about changing rates of deforestation in 16 tropical countries. Deforestation is continuing in all those countries. It is increasing at an alarming rate in Indonesia and Peru. However, in seven countries, the rate of deforestation has declined. Brazil, the largest owner of tropical rainforest, is one of them.

The first concern is the loss of biodiversity. Biodiversity is in itself a natural resource. It is a vital part of the Earth's life-support system. Humans depend on biodiversity for many things that are part of everyday life. Biodiversity provides us with a whole range of goods and services. Some of these have been mentioned in Part 5.2. They range from the food we eat (both animal and vegetable) to timber, from drugs and medicines to fibres for clothing and resins

▶ **Figure 7.20: Changes in annual deforestation rates (2000-2010)**

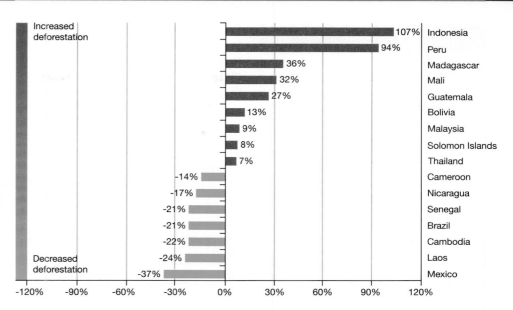

SKILLS ▶ INTERPRETATION (STATISTICAL)

ACTIVITY

Look at Figure 7.20. Explain why the green bars are not necessarily good news.

for glues. Without biodiversity, we would soon perish. Maintaining it is vital to the health of our living planet, and the people who occupy it.

The second, and perhaps even greater concern, is that deforestation contributes to the build-up of greenhouse gases. This, in turn, is generally thought to be a cause of global warming (see Part 7.3). The world's forests absorb carbon dioxide in the atmosphere. Deforestation means that levels of carbon dioxide build up in the atmosphere. Carbon dioxide is not only a major greenhouse gas, it is a potential killer of humans.

▶ **Figure 7.21: Wholesale deforestation in Madagascar – the area was once covered by rainforest**

The two consequences of deforestation just looked at are both highly negative. They are definitely costs. However, deforestation is not necessarily seen in a negative way in all countries.

ACTIVITY

Add another column to Table 7.2 to show the present extent of forest cover as a percentage of the original cover. Write a brief report on what you find.

For a number of developing countries, deforestation can stimulate much-needed economic development. There is no doubt that the Brazilian economy has benefited from exploiting and selling the resources of its vast area of rainforest. Its small neighbour, Guyana, is just beginning to do so. Asian countries such as Indonesia, Thailand and the Philippines have been doing so for longer. But the benefits are only short-term benefits unless something is done to replace the forests that are being cleared. In the Ivory Coast, deforestation of this small West African country is almost complete (Table 7.2). Madagascar is clearing its forests at a swift rate, largely to pay off the country's debts. Are today's Ivory Coast and Madagascar the future for all those countries that still have rainforests?

COUNTRY	ORIGINAL EXTENT OF FOREST COVER (KM²)	PRESENT EXTENT OF FOREST COVER (KM²)	PRESENT RATE OF DEFORESTATION (% PER YEAR)
Brazil	2 86 000	1 800 00	2.3
Colombia	700 000	18 000	2.3
Ecuador	13 200	44 000	4.0
Indonesia	1 220 000	53 000	1.4
Ivory Coast	160 000	4000	15.6
Madagascar	62 000	10 000	8.3
Mexico	400 000	110 000	4.2
Philippines	250 000	8000	5.4
Thailand	435 000	22 000	8.4

▲ Table 7.2: Deforestation of tropical rainforests in selected countries

It is interesting to note that protests about deforestation come most strongly from the governments and people of developed countries. In reply, developing countries ask why should they be denied this chance to develop their economies? How else will they ever catch up with the developed countries? The answer may lie in the sustainable management of the tropical rainforest.

SKILLS   DECISION MAKING

ACTIVITY

Is sustainable management of the tropical rainforest just a dream or a real possibility? Discuss this question in class.

Look back at Part 5.3 (page 134).

7.6 THE IMPACTS OF CLIMATE CHANGE

Although opinion may be divided about the causes of global warming, everyone is agreed that it is taking place. The global pattern of climate is certainly changing. Figure 7.22 shows two inevitable impacts of global warming and where they are most likely to occur on the Earth's surface. While the change in climate will result in some areas becoming drier and others wetter, Figure 7.23 shows two other threats directly related to global warming.

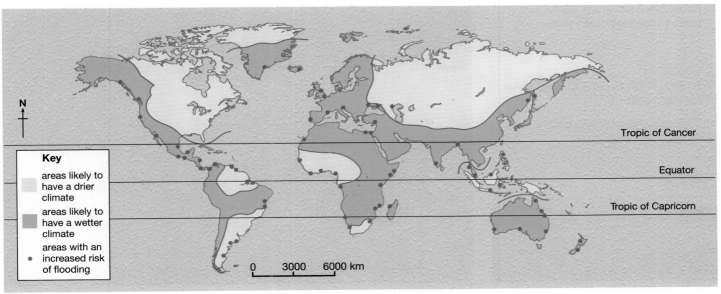

Key

☐ areas likely to have a drier climate

☐ areas likely to have a wetter climate

• areas with an increased risk of flooding

Tropic of Cancer

Equator

Tropic of Capricorn

0   3000   6000 km

▲ Figure 7.22: Possible effects of global warming

▶ Figure 7.23: Challenges posed by global warming

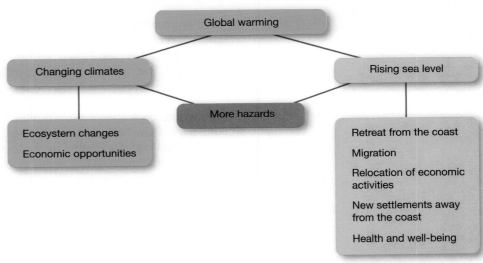

Global warming

Changing climates

More hazards

Rising sea level

Ecosystem changes

Economic opportunities

Retreat from the coast

Migration

Relocation of economic activities

New settlements away from the coast

Health and well-being

MORE HAZARDS

SKILLS DECISION MAKING

ACTIVITY

Make a list of the natural hazards that are expected to increase in frequency and severity with global warming. Which are most likely to affect the country in which you live?

It is generally recognised that another climatic impact of global warming will be to increase extreme weather. More extreme weather means more frequent and more intense natural hazards such as tropical storms, tornadoes, heatwaves, droughts and cold snaps. Warmer seas mean more intense storms (Figure 7.24). Tropical storms have increased in frequency, intensity and power since 1980. Three-day continuous rainfall events in the northern hemisphere, a principal cause of flooding, have also increased significantly since 1980. Global warming is thought to be causing the cyclical ocean current and temperature changes in the Pacific, known as El Nino and La Nina, to become more frequent. These fluctuations in the Pacific have unpredictable global weather effects. With frequent and longer droughts, desertification is increasing and the Sahel region is expanding at an alarming rate. Heatwaves are also on the increase. The persistence of temperatures above 40°C over a period of weeks is thought to have killed 13 000 people in France during August 2003.

▶ **Figure 7.24: A future with more severe tropical storms may look like this.**

▼ **Figure 7.25: The tundra is a biome moving towards the pole.**

ECOSYSTEM CHANGES

The change in climate resulting from global warming will change the distribution of ecosystems. The general, warming will push the world's biomes towards the poles (see Figure 5.1 on page 130). In the northern hemisphere, for example, the coniferous forests will intrude on the tundra, and the tundra on the ice desert (Figure 7.25).

There would be some positive outcomes here. For example, the change in climate will allow farming to be pushed further towards the poles and to higher altitudes. New land will open up for food production in these two locations, as well as in those parts of the world where change will mean a wetter climate. At the same time, however, change in the opposite direction to more arid conditions may result in other areas of farmland becoming less productive.

The retreat of ice in high latitudes will make resources such as oil, natural gas and minerals accessible and allow them to be exploited (Figure 7.26). Much of the Arctic Ocean will become ice-free in the summer and open up the possibility of shipping routes serving the polar coasts of North America and Eurasia.

▶ **Figure 7.26: Oil exploration will push towards the North Pole.**

RISING SEA LEVELS

Higher temperatures have already led to the shrinking of many of the world's major glaciers. In Greenland and the Arctic, ice is melting at a rapid rate and the extent of Arctic pack-ice is shrinking (Figure 7.27). In Antarctica, especially around the peninsula, large sections of ice sheet are breaking off as temperatures rise. As the stores of ice in polar and high mountain regions are unlocked and melt, vast quantities of meltwater will cause the global sea level to rise.

▶ **Figure 7.27: Melting of the Greenland ice sheet**

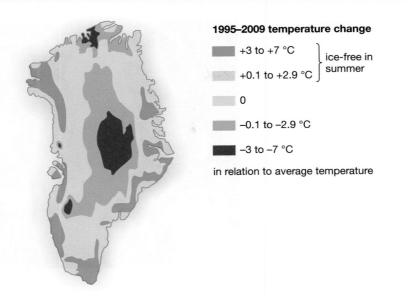

1995–2009 temperature change

- +3 to +7 °C ⎫ ice-free in
- +0.1 to +2.9 °C ⎬ summer
- 0
- −0.1 to −2.9 °C
- −3 to −7 °C

in relation to average temperature

DID YOU KNOW?

Figure 7.27 shows a curious feature. While the edge of the ice sheet is melting and retreating, the core is becoming colder.

▶ **Figure 7.28: Global sea level rise since 1880**

SKILLS ▶ INTERPRETATION (STATISTICAL)

ACTIVITY

Compare Figure 7.14 (on page 193) with Figure 7.28. Do you think that all the graph lines are 'giving the same picture'?

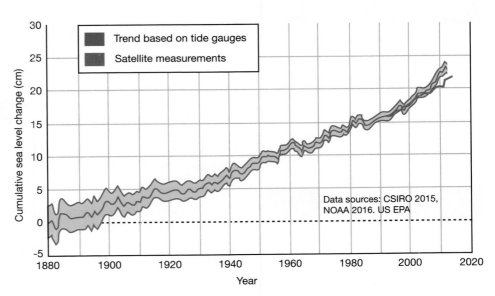

Figure 7.28 shows that the global sea level has risen by 20 cm since 1880. Although there is general agreement that sea levels will continue to rise, there is less agreement about by how much. Most predictions lie in the range of 1 to 2 metres by the year 2100.

Perhaps the greatest threat resulting from climate warming is the rise in the global sea level. Increased flooding and erosion will undoubtedly lead to a retreating coastline, This, in turn, will be followed by some very serious effects.

DROWNING CITIES

The greatest threat is to the world's major coastal cities. Miami (USA) tops the list of the most threatened cities in the world, as measured by the value of property threatened by a 1-metre rise in sea level (Figure 7.29).

This would flood all of Miami Beach and leave downtown Miami on an island of water, disconnected from the rest of Florida. Other threatened US cities include New York, New Orleans, Boston, Washington, Philadelphia, Tampa and San Francisco. Osaka, Kobe, Tokyo, Rotterdam, Amsterdam and Nagoya are among the most threatened major cities outside North America.

CHECK YOUR UNDERSTANDING

Can you name any more leading cities that are at risk?

▶ Figure 7.29: Miami – an endangered city

CHANGING SETTLEMENT PATTERNS

With such a threat in the future, the continued development of low-lying areas around the world has to stop. It would be irresponsible to continue. However, the threat of a rising sea level does not end with the drowning of coastal areas and low-lying islands. What would happen to the people who had to be evacuated? Where would they be resettled? There would be massive volumes of migration. What would happen to all the economic activities and wealth that once occupied the drowned coastal space? Could they be relocated as easily as people? Coastal submergence would have a devastating impact on the economies of many countries, including the USA. The retreat from the coast would cause a major global upheaval. The human settlement pattern would be changed almost beyond recognition.

REDUCED EMPLOYMENT OPPORTUNITIES

Coastal areas contain a large proportion of the world's urban population. But a significant amount of the world's economic wealth is generated and located here as well. A coastal location is important in terms of the global movement of raw materials and manufactured goods.

The need to retreat from the coast will encourage migration. But it will also have a huge negative impact on employment as coastal farmland, cities, tourist resorts and ports are abandoned to the rising sea. It is unlikely that the lost jobs and economic opportunities can be wholly replaced in other locations.

HEALTH AND WELL-BEING

A warmer climate is expected to change the distributions of many diseases. For example, malaria will occur in higher latitudes and at higher altitudes. In those parts of the world that become drier, water will become a scarcer resource. Where this happens, and people are forced to use unclean water, outbreaks of water-borne diseases, such as cholera, typhoid and bilharzia, could become frequent and widespread. Mortality rates would then rise.

Human well-being will be adversely affected, and not just by this raised risk of disease. The upheaval and movement of people and economic activities from coastal areas will cause much human distress and personal insecurity.

CONFLICT

Global warming is expected to increase food and water insecurity. This could easily lead to conflict as people are forced to migrate in search of food and water. It is even possible that people would fight over the remaining sources of food or water. The outlook is bleak.

ACTIVITY

Which countries are likely to win any water wars?

7.7 DEALING WITH DESERTIFICATION

When it comes to dealing with desertification, it is futile to think that its natural causes can be changed. The best that can be done is to help people cope with the natural changes so that the desertification is not made worse.

- Education must be top of the 'action' list. People need to be told of simple actions they can take to make land use sustainable. This is the only way if they wish to survive and remain where they are.
- Reducing soil erosion – this is best achieved by planting trees and perennials (Figure 7.30). Maintaining permanent plant cover ensures that soil is protected from the wind. The cover also helps to maintain soil moisture.
- Changing farming – there is not a lot that can be done here. A system of **permaculture** promises better food security (Figure 7.31). The discovery of certain types of rock that can be used as a fertiliser is possibly promising when growing vegetables.
- Fuelwood – there are two possible actions here: cultivate fast-growing trees for harvesting as fuelwood, and use alternative forms of energy (such as oil) for cooking. The first option might be helped by Relief Everywhere (CARE), an international scheme for growing trees for fuel. However, growing trees means the loss of land for growing food. The second option is probably ruled out by cost and availability.
- Water conservation – it is hard to conserve water if an area receives very little rain over a period of years. For valley areas, there is some hope in a relatively new but simple technology: water-spreading weirs.

Water-spreading weirs are stone and cement constructions that extend from one side of a valley to the other. They are usually built in a series along the length of the valley. The weirs hold back the flow of a river, but allow water to overflow and so pass on downstream to the next weir. Because the speed of water flow is reduced, soil is deposited. Water seeps into the ground, raises the water table and extends the cultivable area. These weirs allow farmers to grow crops all year round (rain-fed and irrigated), even on formerly degraded lands. So, food security and resilience are increased.

Of course, there is the cost of building the weirs. The cost ranges between $US400 and $2000 per hectare of rehabilitated farmland. But the cost is much lower than that of a dam. Micro-loans may be needed to meet the costs, but these are often made available by aid organisations. The extra crop yields help to pay back the loans. The success of the weirs depends a lot on the cooperation of communities strung out along the valley.

▼ Figure 7.30: Terracing and tree planting to limit soil erosion on slopes

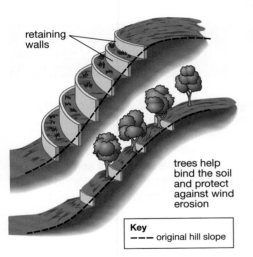

retaining walls

trees help bind the soil and protect against wind erosion

Key
--- original hill slope

SKILLS PROBLEM SOLVING

ACTIVITY

Try ranking these 'ways of dealing with desertification' according to their effectiveness against desertification.

▶ **Figure 7.31: A permaculture plot**

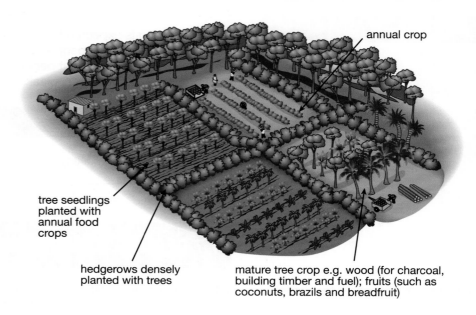

annual crop

tree seedlings planted with annual food crops

hedgerows densely planted with trees

mature tree crop e.g. wood (for charcoal, building timber and fuel); fruits (such as coconuts, brazils and breadfruit)

7.8 MANAGING RAINFORESTS IN A SUSTAINABLE WAY

SKILLS > CRITICAL THINKING

ACTIVITY

Give an example of what you would consider to be appropriate technology in managing the rainforest.

The loss of forests, not only in the tropics, but all over the world is causing concern. Many governments and international organisations recognise the need to manage forests to ensure that resources are there for future generations. This is called sustainable management. The key to sustainability is using resources now in such a way that future generations will still be able to use the same resources. Sustainability is the ability of one generation to hand over to the next at least the same amount of resources that it started with.

Sustainable management of any resource should:

- respect the environment and cultures of local peoples
- use traditional skills and knowledge
- give people control over their land and lives
- use appropriate technology – machines and equipment that are cheap, easy to use and do not harm the environment
- generate income for local communities – not transnational companies
- protect biodiversity.

Figure 7.32 shows how the tropical rainforest may be used in a sustainable way. This may look very easy. But it is difficult for some governments, especially in developing countries, because they need the money that forest exploitation brings.

The sustainable management of forests can be achieved by different methods:

- protection of forests – in some countries areas of forest are conserved and protected as national parks where no or very little development is allowed to take place
- carefully planned and controlled logging in forests
- selective logging of only those trees that are valuable, leaving the rest untouched; for example, in parts of Indonesia only 7 to 12 trees per hectare are allowed to be felled

DID YOU KNOW?

For more information about alternative energy sources, see Part 4.8 (page 118).

- replanting forested areas that have been felled
- restricting the number of logging licences to reduce the amount of forest loss
- heli-logging – for example, in Sarawak helicopters are used to remove the logs because less damage is done to the remaining forest
- developing alternative energy supplies – for example, use of biogas, solar and wind power reduces the amount of wood needed for fuel.

▼ Figure 7.32: Using the rainforest in a sustainable way

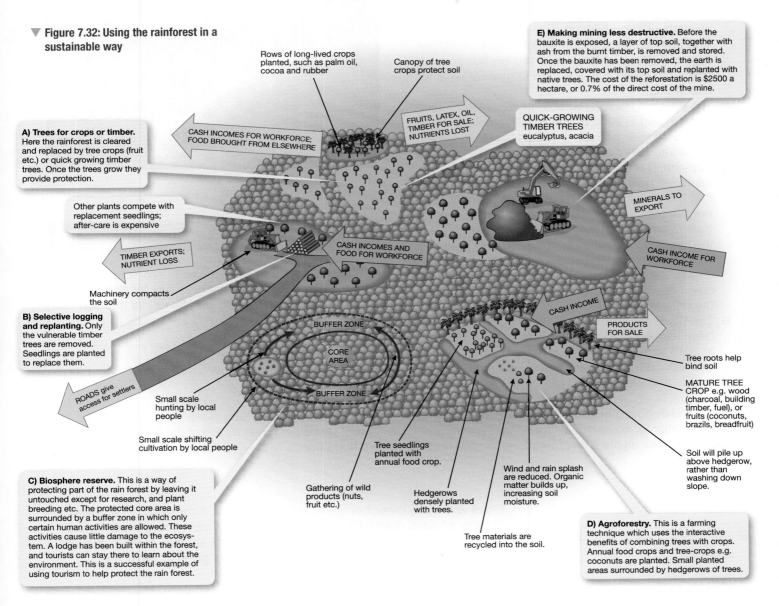

E) Making mining less destructive. Before the bauxite is exposed, a layer of top soil, together with ash from the burnt timber, is removed and stored. Once the bauxite has been removed, the earth is replaced, covered with its top soil and replanted with native trees. The cost of the reforestation is $2500 a hectare, or 0.7% of the direct cost of the mine.

A) Trees for crops or timber. Here the rainforest is cleared and replaced by tree crops (fruit etc.) or quick growing timber trees. Once the trees grow they provide protection.

Other plants compete with replacement seedlings; after-care is expensive

B) Selective logging and replanting. Only the vulnerable timber trees are removed. Seedlings are planted to replace them.

Machinery compacts the soil

C) Biosphere reserve. This is a way of protecting part of the rain forest by leaving it untouched except for research, and plant breeding etc. The protected core area is surrounded by a buffer zone in which only certain human activities are allowed. These activities cause little damage to the ecosystem. A lodge has been built within the forest, and tourists can stay there to learn about the environment. This is a successful example of using tourism to help protect the rain forest.

D) Agroforestry. This is a farming technique which uses the interactive benefits of combining trees with crops. Annual food crops and tree-crops e.g. coconuts are planted. Small planted areas surrounded by hedgerows of trees.

Rows of long-lived crops planted, such as palm oil, cocoa and rubber

Canopy of tree crops protect soil

QUICK-GROWING TIMBER TREES eucalyptus, acacia

FRUITS, LATEX, OIL, TIMBER FOR SALE; NUTRIENTS LOST

CASH INCOMES FOR WORKFORCE; FOOD BROUGHT FROM ELSEWHERE

MINERALS TO EXPORT

TIMBER EXPORTS; NUTRIENT LOSS

CASH INCOMES AND FOOD FOR WORKFORCE

CASH INCOME FOR WORKFORCE

CASH INCOME

PRODUCTS FOR SALE

ROADS give access for settlers

BUFFER ZONE

CORE AREA

BUFFER ZONE

Small scale hunting by local people

Small scale shifting cultivation by local people

Gathering of wild products (nuts, fruit etc.)

Tree seedlings planted with annual food crop.

Hedgerows densely planted with trees.

Wind and rain splash are reduced. Organic matter builds up, increasing soil moisture.

Tree materials are recycled into the soil.

Tree roots help bind soil

MATURE TREE CROP e.g. wood (charcoal, building timber, fuel), or fruits (coconuts, brazils, breadfruit)

Soil will pile up above hedgerow, rather than washing down slope.

Two rather different but related ways of conserving the rainforest are shown below.

- Agro-forestry – combining crops and trees, either by allowing crops to be grown in carefully controlled areas within the forest; or by growing trees on farms outside the rainforest.
- Substitution – finding alternative sources for the resources being taken from the rainforest. The second form of agro-forestry is one example – growing the timber and fuel on plantations elsewhere, and using faster-growing trees. Making better use of savanna areas for livestock grazing is another example.

CASE STUDY: BRAZIL'S SEARCH FOR WAYS TO CONSERVE ITS RAINFOREST

Significant ways to conserve the rainforest include:

- The Forest Code – the code was first created in 1965. It is a law that requires all landowners in the Amazon to maintain up to 80 per cent of their property under rainforest. This means that farmers who buy rainforest can only clear and farm 20 per cent of it. The problem is that the law has proved difficult to enforce as it is hard to monitor what is happening.
- The Amazon Region Protected Areas (ARPA) scheme was created in 2002 by the Brazilian government. Over a 10-year period, 45 million hectares were made into parks and reserves.
- Replanting projects – a project in the Atlantic rainforest near Rio de Janeiro (REGUA) has shown that it is possible to recreate rainforest cover almost like the original. This is done by collecting seeds from remaining patches of primary forest, growing the seeds into saplings in nurseries, and then replanting the saplings back in the deforested areas (Figure 7.33). It is amazing how quickly a new forest cover develops with almost the same gene bank as the original cover.
- The US-Brazil Partnership – it is hoped that what has been learned at REGUA will help Brazil's promise in 2015 to restore 12 million hectares of land now abandoned by commercial farmers because of fertility and productivity.

The Brazilian government has also been encouraging the cultivation of commercial timber (fast-growing species like eucalyptus) on plantations located outside

▼ Figure 7.33: The REGUA nursery produces 70 000 saplings of 180 rainforest species a year.

SKILLS   CRITICAL THINKING

ACTIVITY

Imagine you have been asked to promote replanting projects as the best way for Brazil to conserve its forests. Outline the points that you would make.

the rainforest. There are now 5 million hectares of forest plantations.

All these actions are steps in the right direction, but the task of preventing destruction of the Brazilian rainforest is a huge one.

▼ Table 7.3: International action on deforestation

Brazil and many other countries are realising that, if the tropical rainforest is to have a future, concerted international action is vital. The United Nations has a number of international programmes. There are also some important international treaties. Some examples of both are given in Table 7.3.

INTERNATIONAL PROGRAMMES	AIMS
United Nations Programme on Reducing Emissions from Deforestation and Forest Degradation (UN-REDD)	This programme deals with a wide range of pressing issues, including how best to counter the forces driving deforestation, and how best to ensure the needs of local and indigenous peoples. It also includes making payments if forested areas are left untouched.
United Nations Forum on Forests (UNFF)	The main objective is to promote the management, conservation and sustainable development of all types of forests and to strengthen the long-term commitment of governments to this end.

INTERNATIONAL TREATIES	AIMS
Convention on International Trade in Endangered Species (CITES)	This is an international agreement between governments. Its aim is to ensure that international trade in specimens of wild animals and plants does not threaten their survival.
International Tropical Timber Agreement (ITTA) (2006)	This aims: a) to promote the expansion and diversification of international trade in tropical timber from sustainably managed and legally harvested forests, and b) to promote the sustainable management of tropical timber-producing forests.

One of the biggest challenges for the international community is to stop the huge amount of illegal logging still occurring in the rainforests. Given the remoteness of rainforest areas, illegal logging can easily go unnoticed. Spotting where it is happening can be difficult. However, satellites now help to monitor this. Non-government organisations such as Greenpeace and the Worldwide Fund for Nature (WWF) also play a part in tracking down illegal loggers, as does education. Lobbying organisations running websites such as www.Illegal-logging.info also help.

7.9 RESPONSES TO GLOBAL WARMING AND CLIMATE CHANGE

How a country responds to global warming and climate change is a government decision. Organisations and individuals may support that decision and cooperate. Or they might try to change the official response.

There are four basic responses to global warming and climate change that every country should make:

- join the global effort to reduce greenhouse gas (GHG) emissions
- protect and conserve forests
- identify environments that will become more fragile as a result of climate change; and then decide what, if anything, should be done
- identify coastal areas that will be threatened by flooding or marine erosion and decide what action needs to be taken.

GLOBAL ACTION

Reducing carbon emissions is a global responsibility. Every country needs to be involved, particularly the major polluters (Table 7.4).

At a heads-of-government meeting held in Kyoto (Japan) in 1997, a treaty was drawn up calling for all countries to cut their greenhouse gas emissions by an average of 5 per cent by 2012 (based on 1990 figures). It was thought that this would reduce the rate of global warming. The Paris Agreement (2015) represents the latest attempt to control greenhouse gas emissions. It sets out a global action plan aimed at limiting global warming to well below 2°C above pre-industrial levels. Clearly, this will require drastic action by the leading polluters. The agreement will only come into force when 55 countries, which

▼ Table 7.4: The world's top 10 producers of GHG emissions

COUNTRY	GHG EMISSIONS (MT)	% OF GLOBAL TOTAL
China	9679	22.7
USA	6669	15.6
India	2432	5.7
Russia	2291	5.4
Japan	1257	2.9
Brazil	1105	2.6
Germany	904	2.1
Canada	710	1.7
Iran	698	1.6
Mexico	682	1.6

together account for 55 per cent of global emissions, have signed up and declared their targets. In 2016, both China and the USA signed up, and also the UK.

The following case studies look at the responses of governments at three different points along the development pathway: the UK, China and Sri Lanka.

ACTIVITY

What happens to the national rankings in Table 7.4 if population size is taken into account? Find the population data you need, do the calculations and write a short report on your results.

CASE STUDY: THE UK'S RESPONSE

GHG emissions – the UK is currently responsible for 1.4 per cent of global GHG emissions. It appears very willing to play its part in any global programmes to do with cutting emissions. It signed up to the Kyoto Protocol in 1995. Its own Climate Change Act (2008) commits it to cutting emissions by at least 80 per cent from 1990 levels. Now there is even talk of setting a target of zero carbon emissions.

In order to cut its emissions, the UK has tried to reduce:

■ its reliance on fossil fuels as a primary energy source; that now stands at 85 per cent, but today it is burning much more gas (a cleaner-burning fuel) and far less coal
■ its overall energy consumption by using energy more efficiently; consumption has fallen by 18 per cent since 2005.

The carbon capture and storage option (see Part 4.9) has been ruled out at the moment on the grounds of cost and current technology.

Forests – the UK is the second least-wooded country in Europe. Only 12 per cent of the country is now covered by woodland. That percentage was much higher in the past. There are no plans to reforest on a large scale, but there are now tight controls on imports of hardwood. This is to discourage illegal logging in particular, and felling of the tropical rainforest in general.

▼ Figure 7.34: Dealing with the coastal threat is a UK priority.

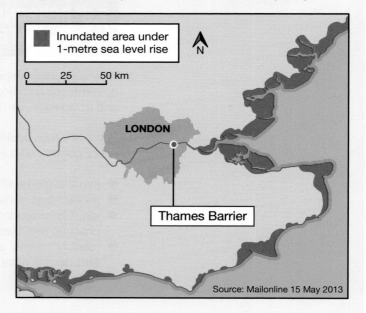

SKILLS CRITICAL THINKING

ACTIVITY

Can you think of any reasons why global warming is not expected to increase the fragility of any of the UK's environments?

Fragile environments – no environments have yet been flagged as being too threatened by global warming.

Coastal risks – these risks are considerable because of the UK's very long coastline and the low-lying nature of much of that coastline (Figure 7.34). As was described in Part 2.9, the coastline has been mapped and classified into those stretches that will be protected from sea-level rises and those that will be left to drown.

It seems that the UK has already responded to the threats linked to global warming and climate change. It has also taken some action to reduce the causes.

SKILLS   PROBLEM SOLVING

ACTIVITY

Would you be in favour of allowing all the areas shown in Figure 7.34 to be flooded by the sea?

CASE STUDY: CHINA'S RESPONSE

Greenhouse gas emissions – China is in a predicament. It is responsible for 22.7 per cent of global GHG emissions (Figure 7.35). Nearly 80 per cent of its energy comes from coal. It is reluctant to turn to oil and gas because it has such large reserves of coal within its borders. The problem is made worse by the fact that China's economic success as an emerging nation has greatly increased the demand for, and consumption of, energy. Heavy industries in China burn large quantities of coal. China adopted the Paris Agreement in 2016. It has also produced its own National Action Plan on Climate Change. This involves a set of different actions that together should make China's economic development more sustainable. But the list of actions does not include either targeting energy or industrialisation.

Forests – 18 per cent of China is covered by forests. Clearly, that figure was much higher in the past. Much of the subtropical and temperate rainforest has been cleared and made into fields. But recently China has started to implement the largest forest conservation and restoration programme in the world. It is claimed this will mitigate (compensate for) some of China's huge emissions of carbon.

Fragile environments – large areas of China frequently suffer from droughts and associated famines. There are also areas prone to widespread flooding. China has recorded six of the world's 10 deadliest floods. It is thought that climate change will make these hazards more severe and more frequent. It would be a huge advantage if China could transfer floodwater to its drought-stricken areas. The South-North Water Transfer project (see Part 1.8) may be a first step in this direction.

Coastal risks – a recent study has found that 50 million people in China will be at risk from coastal flooding

▼ Figure 7.35: Reducing carbon emissions is China's top priority.

SKILLS   REASONING

ACTIVITY

Explain why China was reluctant to sign the Paris Agreement.

SKILLS   DECISION MAKING

ACTIVITY

Will the South-North Water Transfer Project be able to do much to protect those environments made fragile by more droughts and floods?

by the end of this century if GHG emissions stay high. Coastal erosion and shoreline retreat are already serious threats and will only worsen. Coastal zone management is now under way. Prevention measures being implemented include the building of sea walls, land-use zoning and restrictions on land reclamation.

The huge scale of China (its land area, population and economy) means that any response it makes to climate change will make a difference. It will also be very obvious to the global community if it does not respond. At present, the Chinese government is reluctant to do anything that might slow the rate of economic growth but it will need to take a proactive approach to climate change if it is to protect its fragile environments and coastal risks.

CASE STUDY: SOLOMON ISLANDS' RESPONSE

Solomon Islands is a sovereign country consisting of six main islands and over 900 smaller ones located in the Pacific Ocean to the east of Papua New Guinea. It was formerly a British colony, but has been an independent state since 1976. Three-quarters of the labour force are engaged in subsistence farming and fishing. Its main exports are timber, copra, palm oil and small amounts of gold. Tourism is another source of income and employment, but its growth is held back by a lack of infrastructure. The population is just over 600 000.

Carbon emissions from the Solomon Islands are extremely low. They come mainly from the burning of fuelwood in the home, and of oil by motor vehicles and inter-island transport. Tropical rainforests once richly covered the islands, but this is not so today. The export of hardwood timber and the need to create more farmland to replace areas lost by coastal erosion has led to much deforestation.

Fragile environments are being created by several active volcanoes, as well as the fairly frequent earthquakes and tsunamis. Clear felling of the forested volcanic slopes has led to landslides and soil erosion. But the most serious threat is the erosion of the all-too-scarce coastal lowlands, aggravated by a rising sea level. Recently, five tiny uninhabited islands have disappeared below a rising sea level; and six more islands have had great swathes of land and villages washed out to sea.

The Solomon Islands is all too aware of the risks associated with climate change. They know what needs to be done to reduce it, but it has to be done elsewhere and by other countries. Even if there was something they could do, the Solomon Islands simply does not have the financial resources and technology (Figure 7.36). In many respects, people living in the Solomon Islands are the innocent victims of global warming. Their only possible response is to cling to the islands so long as they remain above sea level. In September 2015, they submitted a new climate action plan to the UN Convention on Climate Change. But will that save them?

▲ Figure 7.36: Flooding threatens the Solomon Islands; this man is standing at the location of the original shoreline.

ACTIVITY

Why was it easy for the Solomon Islands to sign the Paris Agreement?

SKILLS ▷ DECISION MAKING

ACTIVITY

Of the three countries – the UK, China and the Solomon Islands – which do you think is most threatened by global warming? Give your reasons.

These case studies show that climate change presents all three countries with considerable challenges. There is a critical need for changes to be made in each country. To date, there has been some response, but possibly not enough. Of all the consequences of climate change, the rise in global sea level is the most serious. In this respect, land-bound countries might consider themselves fortunate.

CHAPTER QUESTIONS

END OF CHAPTER CHECKOUT

INTEGRATED SKILLS
During your work on the content of this chapter, you are expected to have practised in class the following:

- using world maps to show the locations of fragile environments
- using and interpreting line graphs showing population and resources relationships
- using maps to identify patterns of desertifi cation
- using and interpreting maps and graphs relating to human causes of climate change
- using and interpreting line graphs and bar charts showing climate change and sea-level change.

SHORT RESPONSE

1 Define the term 'desertification'. [2]

2 State **one** natural cause of climate change. [1]

3 Identify **one** good and **one** service provided by the tropical rainforest. [2]

4 State **one** way of reducing carbon dioxide emissions. [1]

LONGER RESPONSE

5 Explain why global warming is expected to result in more hazards. [6]

6 Evaluate **three** ways in which the tropical rainforests can be managed sustainably. [6]

7 Explain how desertification affects people. [12]

8 Suggest what makes rising sea levels a threat to people and the environment. [12]

EXAM-STYLE PRACTICE

1 a) Identify **one** greenhouse gas. [1]
   A Hydrogen peroxide
   B Oxygen
   C Carbon dioxide
   D Helium

   b) State **one** type of economic activity that is producing large amounts of greenhouse gases. [1]

2 a) Name **one** feature of desertification. [1]

   b) Study Figure 7.3 (on page 186). Name **two** countries at risk from desertification. [2]

   c) Suggest **two** reasons for this increasing risk. [2]

3 Explain **two** ways in which people are contributing to desertification. [6]

4 a) Study Table 7.4 (on page 208). Calculate what percentage of global greenhouse gas emissions is accounted for by China and India. [2]

   b) Suggest **two** reasons why China and India produce such large emissions of greenhouse gases. [2]

5 Assess the environmental impacts of deforestation. [6]

6 Discuss why some emerging countries are finding it difficult to sign international agreements about reducing carbon emissions. [12]

[Exam-style practice total 35 marks]

8 GLOBALISATION AND MIGRATION

LEARNING OBJECTIVES

By the end of this chapter, you should know:

- The nature of the global economy and the factors encouraging it

- The identity and role of major players in the global economy

- The factors responsible for population migration

- The benefits and costs to countries hosting transnational corporations (TNCs)

- The impacts of migration on different groups of people

- The impacts of the growth of global tourism

- The geological relationships between countries

- Different approaches to the management of migration

- Different approaches to making global tourism more sustainable

This chapter is about one of the most important processes affecting the world today (globalisation) and one of its main outcomes (migration). It starts by looking at the factors encouraging the growth of the global economy, and at some of the major players in that economy. The push and pull factors responsible for today's huge migration flows are examined. Attention then turns to the impacts of three components of globalisation, namely the TNCs, the migration flows and global tourism. The final part looks at the management of migration flows and global tourism by countries at different levels of development.

8.1 THE RISE OF THE GLOBAL ECONOMY

THE NATURE OF GLOBALISATION

DID YOU KNOW?

It would be more accurate to use the term 'economic globalisation' for what has been described in this first paragraph. Many would say that globalisation is also a cultural and political process.

The term globalisation was first used in the 1960s, but it is only since the 1990s that its use has become widespread. It is the process by which the countries of the world are being gradually drawn together into a single global economy. This is a result of the growing network of economic, communication and transport links. Countries are becoming increasingly dependent on each other. This means that economic decisions and economic activities in one part of the world can have important effects on what happens in other parts of the world. Those decisions are being made by the more powerful countries and huge business empires.

▶ Figure 8.1: The growth of the global economy

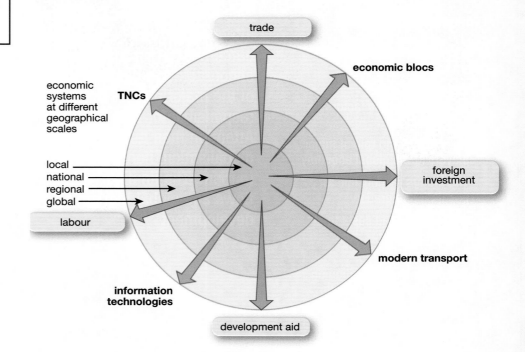

CAUSES OF GLOBALISATION

It could be argued that globalisation is not really a new process. For centuries, countries have been trading with each other, often over long distances. The colonial empires of Britain, France and Spain built up during the 16th, 17th and 18th centuries could be described as early expressions of globalisation. What has changed since the middle of the 20th century is the scale of international trading and of the other economic links that are creating this interdependence (Figure 8.1).

Four significant developments have helped this change.

■ The appearance of large transnational corporations (TNCs) with diverse business interests spread across the globe.
■ The growth of regional economic or **trading blocs**, such as the European Union (EU) and the North American Free Trade Agreement (NAFTA). By encouraging free trade between member countries, the barrier effects of national boundaries are broken down. So, there is much more global trade.

The development of modern transport networks (air, land and sea) capable of moving people and commodities quickly and relatively cheaply. Due mainly to aircraft, physical distances worldwide are now much less important. We live in a 'shrinking world'.

Advances in communications and **information technology** mean that important data and decisions can be whizzed around the globe in a matter of seconds. A TNC with its headquarters in London or another major city can closely monitor market trends around the world. It can easily check on what is happening in its branch offices and factories scattered worldwide. Decisions can be quickly transmitted.

THE WORKINGS OF THE GLOBAL ECONOMY

The outcome of these developments is today's **global economy**. There is scarcely a country in the world that is not participating in some way or another. The workings of the global economy involve five different forms of flow (Figure 8.1):

- trade – through the export or import of raw materials, food, finished goods or services
- aid – either as a donor or a receiving nation; much aid is of an economic nature
- **foreign investment** – through investment, TNCs are able to exploit economic opportunities around the world, for example oil in west Africa or sugar in Brazil
- labour – this is vital to the workings of the global economy; economic migration is commonplace these days as people move in search of work and a better life, and TNCs are constantly on the look-out for cheap labour
- information – the fast transfer of data and decisions is crucial to the workings of the global economy.

ACTIVITY

On an outline map of the world, plot the production chain shown in Figure 8.2.

▶ **Figure 8.2: The production chain of a pair of jeans**

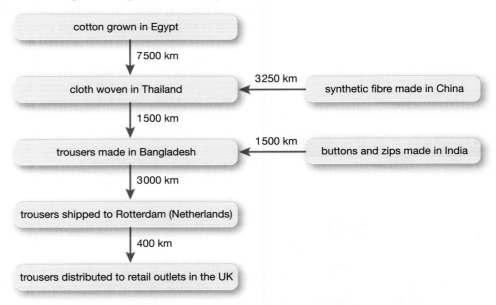

Through the growing global economy and its flows, we are all drawn into a global community of interdependent nations. What is the evidence for this? Most often mentioned are the **production chains** (also known as **commodity chains** or supply chains). A production chain consists of a number of stages involved in the making of a particular product. At each stage, value is added to the emerging product. Figure 8.2 shows the production chain involved in making a pair of jeans. The reasons for companies setting up these transnational production chains are explained in Part 8.2.

▼ Figure 8.3: A busy call centre in India

Other evidence of the workings of the global economy includes:

- the **call centres** for UK and USA companies located in India, Philippines and Thailand (Figure 8.3)
- the **outsourcing** of food production and manufacturing in developing countries where labour is cheap
- the global shift in manufacturing from developed to emerging or developing countries
- the growing volumes of economic migrants and international tourists.

The growth of the global economy depends on advances in transport and communications. The working of the economy involves the four international links: trade, foreign investment, aid and the movement of labour. There is plenty of evidence of the global economy, from modern production chains to the global shift in manufacturing.

SKILLS   REASONING

ACTIVITY

Why are there so many call centres in India, the Philippines and Thailand?

8.2 THE ROLES OF GLOBAL INSTITUTIONS (THE MAJOR PLAYERS)

The working of the global economy involves a number of major players, that is organisations that have great power and influence. They include global organisations such as the United Nations (UN), the World Bank, the International Monetary Fund (IMF) and the World Trade Organization (WTO). Then there are the great business empires known as transnational corporations (TNCs) or multi-national corporations (MNCs). Other powerful players are the USA, the European Union, Japan and OPEC.

GLOBAL ORGANISATIONS

There are two inter-governmental organisations (IGOs) which play a leading role in the global economy.

▲ Figure 8.4: The WTO – the global 'umpire' of fair trade

- International Monetary Fund (IMF) – this was set up in 1944. Its main aim is to ensure that the exchange rates between the currencies of the world remain stable. It also runs a system of international payments. Both these services which involve flows of money are vital to the workings of the global economy.
- World Trade Organization (WTO) – this was set up in 1995. As the name indicates, its aim is to promote free and fair trade between countries and to ensure that every trading nation keeps to the rules of international trade (Figure 8.4). Remember that trade is very important to the workings of the global economy.

These two organisations are major players in global trade and foreign investment. But as Figure 8.1 (page 218) shows, there are other important links in the global economy. One of these is development aid. This focuses on reducing poverty, and promoting education and healthcare. Again, there are important players: one IGO and a number of NGOs.

SKILLS REASONING

ACTIVITY

Explain why the WTO might be thought of as an 'umpire' as suggested in Figure 8.4.

DID YOU KNOW?

Three of the organisations were set up in the mid-1940s. This was as the Second World War was coming to an end. Some governments at that time felt that organisations should be set up to help the world recover from the human disaster.

- World Bank – this IGO was also set up in 1944. It provides low-interest loans and technical help to developing countries. Its aims are to reduce poverty and help development. It supports a wide range of projects from education and health to agriculture and infrastructure. See more in Part 9.7, page 261.
- NGOs – non-governmental organisations such as Cooperative for Assistance and Relief Everywhere (CARE), Oxfam International, Action Against Hunger (AAH) and the Red Cross and Red Crescent. Their efforts are targeted at poverty and hunger.

TRANSNATIONAL CORPORATIONS (TNCS)

The appearance of these vast business empires is a striking feature of globalisation. Some of them are leading players in the global economy. They have great power and influence.

Table 8.1 gives the names of some of the world's leading TNCs. It is interesting to note that over half of them are involved in the oil industry. This reflects the fact that oil and gas are currently the leading sources of global energy. They are vital to the workings of the global economy – as raw material sources, fuel for transport, and as generators of electricity for industry and the home.

▶ Table 8.1: The top 10 TNCs in 2016 (by revenue)

RANK	COMPANY	INDUSTRY
1	Walmart	Retail
2	State Grid	Electricity
3	Samsung	Conglomerate
4	China National Petroleum	Oil and gas
5	Sinopec Group	Oil and gas
6	Royal Dutch Shell	Oil and gas
7	Exxon Mobil	Oil and gas
8	Volkswagen	Automobile
9	Toyota	Automobile
10	Apple	Consumer electronics

CHECK YOUR UNDERSTANDING

Try naming five more well-known TNCs which have their main business interests outside the oil industry.

The production chains of these and other TNCs cross the globe bring the countries of the world together into a network of interdependence (see Figure 8.2 on page 215). The overriding motive for setting up these chains is to maximise sales and profits. Four reasons for this were noted on page 214. There are also other strong reasons:

- to be close to major markets
- to sell inside trade barriers
- to take advantage of incentives offered by governments
- to be able to operate without too many restrictions.

The advantage of being close to major markets is well illustrated by the factory that the Japanese car manufacturer Nissan set up in Sunderland (north-east England) in 1986.

8.3 THE GROWING VOLUME OF MIGRATION

Figure 8.5 provides a snapshot of global population change in 2013. It shows some strong contrasts. Compare the high rates of increase in Africa and the Middle East and the little or no gain in North America, Europe and Russia. In general, the growth rates in Figure 8.5 are the outcome of natural population change – the product of high **birth rates** and falling **death rates**. By comparison, the impact of migration on the distribution of population change is very small.

▶ **Figure 8.5: Global population growth rates (2013)**

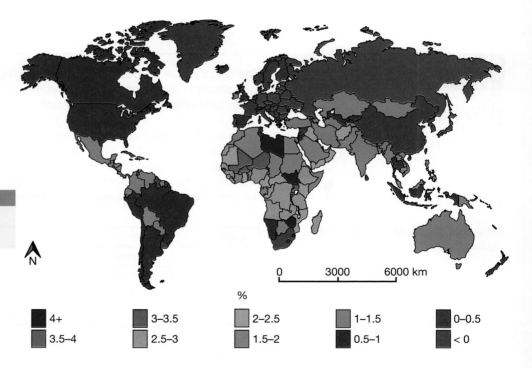

CHECK YOUR UNDERSTANDING

How are birth and death rates calculated?

DID YOU KNOW?

The definition of migrant varies. In the UK and many other countries, a person becomes a migrant if they move to another location and live there for more than a year.

The only type of migration that might make an impact on Figure 8.5 would be international migration – migration from one country to another. Internal migration – the movement of people within a country – will only affect the distribution of population within that country.

Having said that migration's impact on the global population is small compared with that of natural increase, globalisation has encouraged more and more people to migrate. There are a number of different types of migration, which need to be distinguished. We have already mentioned international and internal migration. Another equally significant distinction is that between voluntary (by choice) and forced (compulsory) (Figure 8.6).

▶ **Figure 8.6: A classification of population movements**

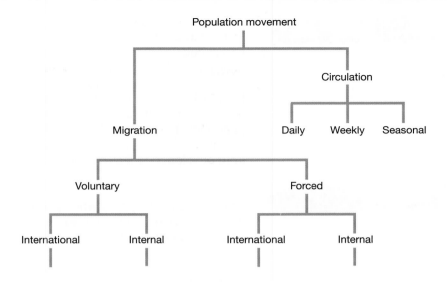

VOLUNTARY MIGRATION

▲ **Figure 8.7: Lionel Messi – a sporting migrant from Argentina to Barcelona, Spain**

Voluntary migration happens when people choose either to move inside their own country or to emigrate to another country. The normal reasons for this are economic (such as to find work or for higher wages) or for a better quality of life. In developing and emerging countries, this usually means moving from rural areas into towns and cities. In developed countries, there is an increasing volume of migration in the opposite direction as people move from crowded large cities into smaller urban settlements or even to the countryside. This is called counter-urbanisation.

One outcome of globalisation has been a huge increase in the volume of voluntary migration, largely for economic reasons. A feature of much of this movement is that it is not migration in the strictest sense. Many of these so-called **economic migrants**, such as those arriving in the UK from Eastern Europe, stay for less than a year. However, some end up staying for a number of years. We can refer to these short-stay workers and their families as 'temporary economic migrants'. A large proportion of the huge numbers of people risking their lives to cross the Aegean and Mediterranean seas into Europe are economic migrants.

In developed countries, a relatively new form of voluntary population movement is retirement migration. This now has an international dimension which can be seen as an outcome of globalisation. For example, UK citizens are retiring to other countries, such as Spain, Portugal, France or even Australia. Another new form of migration involves professional sports people. Increasing numbers, particularly footballers and cricketers, are signing two to five-year contracts to play for a foreign club (Figure 8.7).

SKILLS ▷ SELF DIRECTION

ACTIVITY

List four more 'sporting migrants' and find out their country of birth and where they now live.

CIRCULATION

Figure 8.6 distinguishes between migration (a permanent change of residence) and circulation (a temporary absence from home). Circulation covers any non-residential movement, from a shopping trip and commuting to tourism. Tourism alone gives rise to increasing volumes of population movement (see Part 8.6). Tourism today is not just about taking a break from work and relaxing somewhere. People are travelling abroad for various types of medical treatment. These range from cosmetic and dental surgery to fertility and obesity therapies. What is new about this medical tourism is that the flows are from developed to less developed countries, where treatments are cheaper.

FORCED MIGRATION

Forced migration occurs when people have to move from where they live. They have no choice. Typically, they go to another country, but sometimes they may only be displaced within their own country.

There are many causes of forced migration. Natural hazards such as earthquakes, volcanic eruptions, violent storms, floods and droughts are all physical reasons for having to move. In most cases, the survivors of such disasters will move back home when it is safe to do so or when their homes and jobs are available again.

▶ Figure 8.8: Elderly displaced people take shelter in a school after the 2011 tsunami hit eastern Japan.

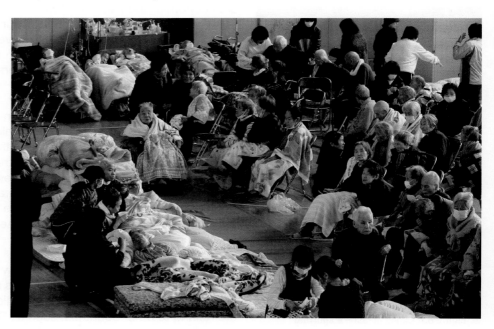

The biggest causes of forced migration result from the actions of people, especially war and persecution. Historically, this includes large-scale migrations; for example, Jews fleeing from the Germans and Russians during the Second World War (1939–1945) and the Palestinian Arabs displaced by the creation of the Jewish state of Israel in 1948.

Many recent wars have been civil wars – factions within a country fighting one another. In some cases, this has been in an effort to force out entire **ethnic groups** or communities – a process known as **ethnic cleansing**. This was the case when the member states of the former Yugoslavia erupted into a series of civil wars in the 1990s. At the same time, in Rwanda, the Hutus attempted to remove the Tutsis, leading to 800 000 deaths and 2 million displaced people. Today, the prime example is the conflict between Sunni and Shia Muslims in the Middle East.

SKILLS ▷ SELF DIRECTION

ACTIVITY

Find out what groups were involved in the ethnic conflict that followed the break-up of the former Yugoslavia.

REFUGEES

CHECK YOUR UNDERSTANDING

Check that you understand the difference between a refugee and an economic migrant.

SKILLS ▶ INTERPRETATION

ACTIVITY

Make brief notes about the global distribution of persons of concern. Find out the reasons for the large number shown for Colombia.

The United Nations High Commission for Refugees (UNHCR) has responsibility of all those people who are forced to migrate. These people are collectively referred to as 'persons of concern' to the UNHCR. Four different categories of person are recognised:

■ **refugee** – a person who, owing to a well-founded fear of being persecuted (on account of their race, religion, political opinion or social group), lives outside their country of nationality.
■ **asylum seeker** – a refugee who has applied to become a citizen of the country where they have sought protection
■ internally-displaced person (IDP) – a person forced to flee their home for the same reasons as a refugee or to escape natural disasters, but they do not cross an internationally recognised border
■ returnee – a refugee or asylum seeker who has voluntarily returned to their own country or an IDP who has returned home.

In 2015, UNHCR recognised that there were 65.3 million forcibly displaced persons in the world. Of these, about one-third were refugees. Figure 8.9 shows the global distribution. The Middle East, Africa and Colombia stand out as the hotspots.

▼ **Figure 8.9: The global distribution of persons of concern (2015)**

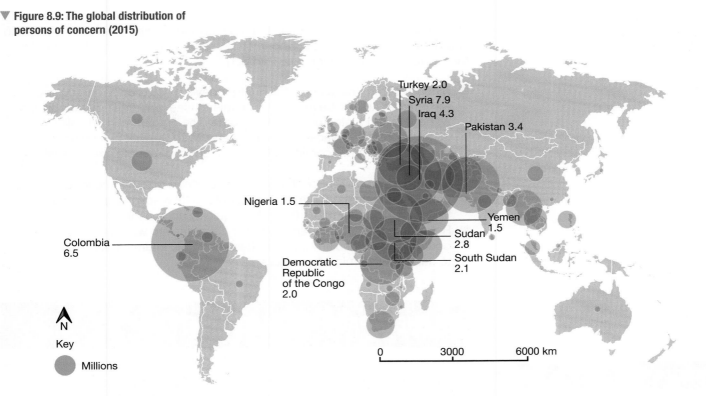

PUSH-PULL MECHANISM

The decision to migrate is usually the outcome of the push-pull factors. Figure 8.10 shows that the push force occurs in the potential migrant's home location. It is something that pushes the person to move away. In the case of forced migration, the push factor is paramount. The pull force is something that attracts the migrant to a particular destination. Very often the pull factor is the mirror image of the push factor. For example, being out of work gets a person thinking that they must move to find a job. They hear that there is a labour shortage in a particular city or country. Therefore the combination of the push and pull factors persuades the person to migrate. In the case of much voluntary migration, the pull factor is stronger in that it often strongly influences the eventual migration destination.

▶ **Figure 8.10: The push-pull mechanism**

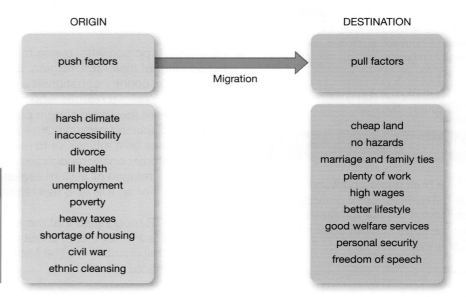

ORIGIN

push factors

Migration

DESTINATION

pull factors

harsh climate
inaccessibility
divorce
ill health
unemployment
poverty
heavy taxes
shortage of housing
civil war
ethnic cleansing

cheap land
no hazards
marriage and family ties
plenty of work
high wages
better lifestyle
good welfare services
personal security
freedom of speech

SKILLS ▶ INTERPRETATION

ACTIVITY

Are you able to add anything more to the lists of push and pull factors in Figure 8.10?

The growing volume of migration in today's world is not just a reaction to push and pull factors becoming stronger. Three other factors that are important in the context of globalisation (see Part 8.1).

■ Modern communications – thanks to today's mass media, particularly the internet, would-be migrants are able to 'see' and 'feel' distant places without taking a step outside their home. The amount and reliability of information about places is much greater. With the risk of moving to an unknown and unwelcoming destination reduced, people are more willing to migrate.
■ Modern transport – once the decision has been taken to move to a particular destination, the migrant is able to take advantage of modern transport. This can move them there quickly and cheaply.
■ Relaxing national boundaries – many countries are willing to relax their boundaries, particularly if it is in their economic interests, for example, to admit skilled migrant workers.

8.4 THE IMPACTS OF GLOBALISATION

GOOD OR BAD?

The growth of globalisation has given rise to a major debate about its real benefits. Its supporters point out that it gives the poorest countries some chances of economic development. They say that even the poorest countries have something to offer to the global economy. Being involved in the global economy creates jobs, the opportunity for people to earn a steady wage and a chance to improve their quality of life (Table 8.2).

▶ Table 8.2: Benefits and costs to countries hosting TNCs

BENEFITS	COSTS
Trade links with other countries set up	Profit-driven
Jobs created; regular wages	Profits 'leak' out of the country
Infrastructure developed	Investment moved to more profitable places
Foreign currency earned from exports	Exploitation of workers
Skills training of local labour	Often little regard for the environment
Investment in new technology	New technology may reduce workforce

CHECK YOUR UNDERSTANDING

Check that you understand the benefits and costs in Table 8.2.

The trouble is that TNCs are businesses which are focused on maximising their profits. They can often be exploitive. Also, they may ignore the environmental and social impacts of their investments. Few TNCs are answerable to the governments of the countries in which they invest. They are so powerful, they can do almost whatever they like. The profits that they make in any one country are generally used to open up new businesses elsewhere. Any investment can disappear as quickly as it came, if global or local economic conditions change.

A VERDICT

The world's poorest countries have yet to see much benefit from globalisation. If anything, it has increased the so-called development gap between the rich and poor nations of the world (see Chapter 9). A similar widening rich-poor divide also applies to the people of countries hosting TNCs. Certainly, TNCs and the countries in which they have their headquarters have become richer. So, too, have corrupt government officials in the host countries. A small number of people have found employment. But how much has the whole population of a host country benefited? What has the TNC done to raise living standards and provide better education and healthcare? The short answer to these questions is little, if anything.

SKILLS ▷ DECISION MAKING

ACTIVITY

In groups, discuss how TNCs might be persuaded to do more for the countries in which they operate.

8.5 THE IMPACTS OF MIGRATION

The possible impacts of migration are almost as numerous as the number of migrants. This is a huge topic. The impacts depend on who the migrant is and their reasons for moving from one location to another. Were they forced to move or was it their choice? We also need to realise that migration moves have an impact on two locations – the one that was left and the one that is the new home.

In order to make this topic manageable, let us take a look at examples of migrants involved in six different types of migration.

CASE STUDY A: VOLUNTARY, INTERNATIONAL

A male aged 30 who has decided to move from Bangladesh to the UK to join some cousins who migrated to Bradford over five years ago.

IMPACT ON MIGRANT
- Needs to adjust to a very different type of society
- Better housing; more to do
- A sense of excitement and possible opportunities
- Becomes aware of ethnic discrimination

IMPACT ON NEW LOCATION
- Finds menial work in a restaurant selling Indian food
- Another person to share a room in his cousin's house
- Adds to the number of Bangladeshis concentrated in a particular part of the city

IMPACT ON OLD LOCATION
- One less person depending on informal employment
- One less person living in the crowded family home
- Remittances are a welcome addition to the family income
- More members of the family may be encouraged to make the move

CASE STUDY B: VOLUNTARY, NATIONAL

A Belgian couple making a retirement move from a city to a coastal resort.

IMPACT ON MIGRANT
- Need to adjust to a different living environment and perhaps a smaller dwelling
- Finding things to do in the new leisure time
- Challenge of making new friends, perhaps joining clubs and societies

IMPACT ON NEW LOCATION
- Adds to the ageing profile of the population
- More strain on healthcare services
- Perhaps adding to the body of voluntary workers

IMPACT ON OLD LOCATION
- A dwelling released for a younger family to occupy
- Friends and family left behind

CASE STUDY C: VOLUNTARY RURAL-URBAN

A young married couple, with one young child, who have decided to leave their rural home in the north of Kenya to find a new life in Nairobi.

IMPACT ON MIGRANT
- Unfamiliarity with the new surroundings and the urban way of life
- Struggle to find housing and a job
- Become victims of urban poverty
- Perhaps feelings of regret that they moved

IMPACT ON NEW LOCATION
- Another family trying to find shelter in a squatter settlement
- Another family on the streets and involved in informal employment

IMPACT ON OLD LOCATION
- Loss of young able-bodied labour
- Adds to the 'ageing' population
- Likely to encourage more people to make the move

CASE STUDY D: FORCED, INTERNATIONAL

A young widow and her 2-year-old daughter who fled the civil war in Syria and who have been accepted as refugees in Germany.

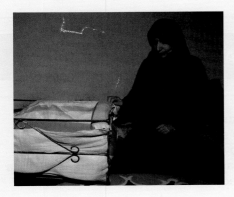

IMPACT ON MIGRANT
- The horrors and hazards of the journey which is a highly risky sea crossing from Turkey to Greece; extortion by people traffickers
- Having to adjust to a very different type of society

IMPACT ON NEW LOCATION
- Two more refugees among hundreds of thousands to be provided with housing and support
- Rising hostility to increasing number of refugees
- Positive impacts for host countries who accept refugees, e.g. increased labour force

IMPACT ON OLD LOCATION
- Family and friends left behind
- Two fewer people at risk in the fighting and bombing
- Decreased labour force.

CASE STUDY E: FORCED, NATIONAL

A family in Ethiopia forced to move by the spread of the Sahel and starvation.

IMPACT ON MIGRANT
- The anxiety of finding a new location where they might settle
- Hunger
- Fatigue of journey on foot in search of a new location

IMPACT ON NEW LOCATION
- Hostile attitudes of food producers
- More mouths to be fed by the same amount of overworked land

IMPACT ON OLD LOCATION
- Reduction in population numbers may create an opportunity for the location to recover some of its fertility

CASE STUDY F: FORCED, RURAL–URBAN

A Chinese family (two adults and one teenage child) whose rural home is one of many thousands about to be flooded to create one of the reservoirs of the SNWTP.

IMPACT ON MIGRANT
- The trauma of being uprooted from home area and losing home
- The loss of the sense of belonging to a local community

IMPACT ON NEW LOCATION
- Another family to find housing in an already overcrowded city

IMPACT ON OLD LOCATION
- The abandonment of another family home and farmstead.

ACTIVITY

Classify each of the following migrants:

- a UK couple retiring to Spain
- a Japanese family whose village had to be abandoned because of damage to a nuclear power station.

8.6 THE IMPACTS OF GLOBAL TOURISM

THE GROWTH OF GLOBAL TOURISM

Globalisation is not just about manufacturing. It is also about services. The UK retailer Tesco, for example, now has supermarkets in many countries in Europe and Asia. But no service has grown and spread more spectacularly over the last 50 years than tourism. Today, around 900 million people become international tourists each year. That is equivalent to more than one-tenth of the world's population.

People have been taking foreign holidays for centuries. Up until the 20th century, however, it was something that only wealthy people did, and often for months at a time. But today's international tourist does not have to be wealthy, and most overseas visits only last for a week or two. Figure 8.11 shows that during the second half of the 20th century, the number of international tourists increased by nearly 30 times.

▼ Figure 8.11: World tourist arrivals, by region (1950-2020)

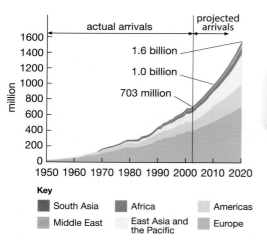

▼ Figure 8.12: The main causes of the rise in tourism

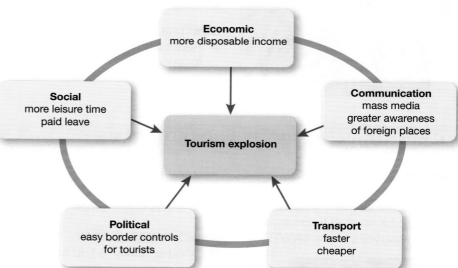

What has caused this change and the spectacular rise in the volume of tourism? There are many reasons (Figure 8.12).

- Most workers in developed countries now work less than 40 hours a week and enjoy up to 6 weeks of paid annual leave (more in Europe than the USA). This combination of more leisure time and paid holidays, plus increasing amounts of disposable income, has given a powerful boost to tourism. Many people are taking early retirement and want to travel.

SKILLS   CRITICAL THINKING

ACTIVITY

Why do people tend to go abroad for their holidays?

CHECK YOUR UNDERSTANDING

Make sure that you understand each of the five causes of the tourism explosion shown in Figure 8.12.

■ Developments in transport have had a big impact on travel. Journey times have been dramatically reduced. Long journeys are now more comfortable, and the relative costs of travel are lower. The introduction of modern forms of transport, such as wide-bodied jet aircraft, hotel cruise ships, cruiser coaches and high-speed rail networks, have all helped to make global travel a reality for increasing numbers of people. Even the world's most remote places are now accessible, for example the Antarctic and Amazon rainforest.

■ The mass media, especially TV and the internet, have increased people's awareness of faraway places and possible tourist destinations. People now know about different destinations, as well as the diversity of holiday activities, from scuba diving and birdwatching to visiting cultural and historic sites.

■ More and more countries are realising the benefits of being a tourist destination, and relaxing their border controls. Governments may make large sums of money from tourist visas and departure taxes. Even the EU, which makes it difficult for workers to enter, warmly welcomes tourists.

■ Tourism has become a 'commodity' which is marketed in much the same way as any new product. There is now a huge business sector made up of travel agents and tour operators set on promoting tourism. One big promotion has been the package holiday. This consists of transport and accommodation, which is advertised and sold together by a tour operator. Other services in the package might include a rental car, special activities and excursions.

THE IMPACTS

The package holiday is the key part of what is known as mass tourism. This is a branch of tourism in which large multinational companies shape developments according to global demand. It is large-scale, highly commercial and focused on popular destinations. It pays little regard to local communities. As with most developments in our modern world, mass tourism has brought both benefits and costs. These positive and negative impacts fall mainly under three headings – economic, socio-cultural and environmental.

▶ Table 8.3: International tourism receipts (2015)

SKILLS   PROBLEM SOLVING

ACTIVITY

What do you think are the attractions that draw so many international tourists to the USA, Spain and China?

RANK	COUNTRY	TOURISM RECEIPTS (US$ BILLION)
1	USA	204.5
2	China	114.1
3	Spain	56.5
4	France	45.9
5	UK	45.5
6	Thailand	44.6
7	Italy	39.4
8	Germany	36.9
9	Hong Kong (China)	36.2
10	Macau (China)	31.3

ECONOMIC

There is no doubt that tourism has positive impacts. Table 8.3 shows the top 10 destination countries in terms of their earnings from international tourism. What the table does not show is the growing number of countries which benefit from tourism in Southern Europe, the Middle East, Africa, South-East Asia and Central America.

▶ **Figure 8.13: The economic multiplier effect of tourism**

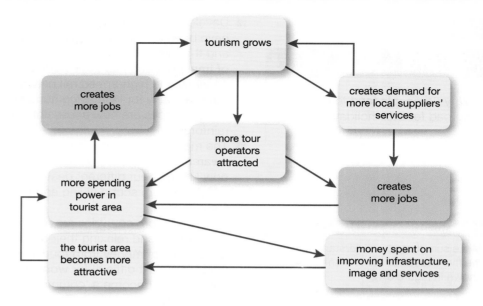

All tourist destinations benefit from tourism's multiplier effects (Figure 8.13). Tourism is labour-intensive and so creates many jobs; not just in hotels and restaurants, but in other tourist services such as transport. While tourism is a service sector activity, it indirectly impacts on the other two sectors of agriculture and manufacturing. Tourists need food, so that is potentially good for agriculture. Tourists buy souvenirs and that can be good for manufacturing. The hotel staff, the ice-cream sellers and the souvenir shop owners then spend their money in the local shops. Tourism puts money into many people's pockets and, through the multiplier effect, the whole local economy can benefit. Few would disagree that tourism can do much to help economic development in developing countries.

However, there are some negative aspects. Much of today's international tourism is in the hands of big companies, such as TUI, Thomas Cook and First Choice. As we saw in Part 8.3, this means that the profits made in a particular country 'leak' out to the country where the tour operator has its head offices. This is money that could be used to help the development of the country where it was earned. For example, in Vanuatu – a 'trendy' destination in the Pacific Ocean – 90 per cent of the profits go to foreign companies. Tourist destinations, particularly in developing countries, can become very dependent on foreign companies.

SOCIAL-CULTURAL

The degree to which tourism impacts on people and their traditional ways of life – their culture – depends on the type and volume of tourism. In some places, mass tourism has helped to revive local handicrafts as well as the performing arts and rituals – if only as a commercial entertainment for visitors. However, generally speaking, the socio-cultural impacts of tourism are negative. The greater the number of tourists converging on a location, the more likely is tension with local people. Tourists can easily offend the traditional values of local people and their codes of behaviour in a number of ways:

- ■ drinking too much alcohol and becoming loud and offensive
- ■ ignoring local dress codes and revealing too much flesh
- ■ encouraging prostitution and, unintentionally, crime
- ■ eroding the local language by relying too much on English
- ■ failing to behave in a proper way in churches, temples and mosques.

SKILLS ▷ CRITICAL THINKING

ACTIVITY

So, how would you sum up the economic benefits of tourism?

SKILLS ▷ CRITICAL THINKING

ACTIVITY

Can you think of any ways of increasing the social-cultural positives?

ACTIVITY

Can you think of any environmental positives associated with tourism?

It would be good to think that international tourism provides the opportunity for people of different cultures to mix and learn about each other. However, such a positive effect happens rarely. Indeed, in some parts of the world tourists are deliberately kept away from local people (or is it the other way round?). This happens, for example, in Cuba (for political reasons) and in the Maldives (for religious reasons).

ENVIRONMENTAL

It is also difficult to identify many positive environmental impacts. Again, it depends on the nature and the volume of tourism. Alternative ecotourism does provide some opportunities for people to learn about the environment and to become supporters of environmental conservation. However, it is easy to compile a long list of negative impacts:

- the clearance of important habitats, such as mangrove and rainforest, to provide building sites for hotels
- the overuse of water resources
- the pollution of the sea, lakes and rivers by rubbish and sewage
- the destruction of coral reefs by snorkelers and scuba divers
- the disturbance of wildlife by safari tourism, hunting and fishing
- traffic congestion, air and noise pollution.

CASE STUDY: SPAIN'S COSTA BLANCA – A PREMIER PACKAGE HOLIDAY DESTINATION

Spain's Costa Blanca is a 200 km stretch of Mediterranean coast running either side of Alicante (Figure 8.14). It is probably the most famous stretch of coast in Spain. In the 1950s it was a fairly quiet coastal area, relying heavily on fishing. However, since the growth of cheap package holidays in the 1960s, the area has been completely transformed. Today it is an almost unbroken strip of high-rise hotels, holiday apartments, shops, cafes and restaurants. Millions of tourists are attracted here by the clear blue waters, the vast white sand beaches, the hot, dry, sunny weather and the wide range of leisure facilities, from water sports to golf courses.

This growth of mass tourism and the package holiday has brought many advantages to this part of Spain, especially in the form of jobs. More recently, the infrastructure of the region has seen a number of improvements, benefiting visitors and local residents alike. This is part of the multiplier effect – more tourists arrive, spending more money which creates jobs, not just in tourism but in the construction industry and for local suppliers, for example farmers.

However, such high numbers of visitors concentrated mainly from May to October, visiting a relatively small area around the Mediterranean coast, has increased pressure on limited resources.

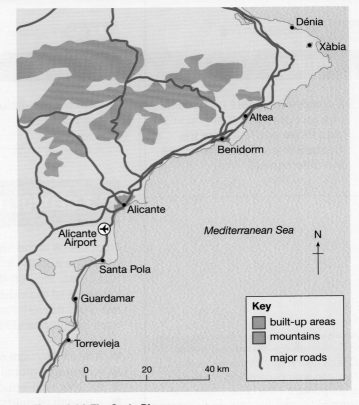

▲ Figure 8.14: The Costa Blanca

Pressures include:

■ high demand for water in areas where it is often a scarce resource; tourists typically use almost twice as much water per day as local residents

■ the production (and disposal) of over 50 million tonnes of waste each year

■ increased urbanisation of coastal regions as more hotels and tourist facilities are built, damaging local ecosystems

■ the increase in the number of second or holiday homes, which take up much more land than hotels but are usually only occupied for short periods

■ high levels of pollution, particularly from cars, aircraft and boats.

Alicante is the capital and major city of Costa Blanca, but it is Benidorm that attracts the most visitors. Benidorm is one of the most famous modern Mediterranean holiday resorts (Figure 8.15). It has a permanent population of about 70 000, but during the peak season the population is more than half a million.

The development of Benidorm as a coastal resort started in 1954, when its young mayor drew up an ambitious plan of urban development. The whole project really took off in the 1960s when it became popular with British tourists on summer package holidays. Today, Benidorm's tourist season is all year round, and its attractions are now much more than 'sun, sea and sand'. The nightlife – based on a central concentration of bars and clubs – is a strong pull, especially among younger people. Benidorm has been transformed from a small sleepy village into a modern pulsating urban area of skyscraper hotels and apartment blocks, theme parks, pubs, clubs and restaurants. It is a tourist hotspot in every sense of the word. But at what cost to a once beautiful stretch of coast and the quality of inshore waters?

▲ Figure 8.15: Benidorm seafront

ACTIVITY

Compare the costs and benefits of a tourist region that you know with those you listed for the Costa Blanca, Spain. What are the differences?

ACTIVITY

Make a two-column table listing the costs and benefits of tourism in this part of Spain.

8.7 GEOPOLITICAL RELATIONSHIPS

Geopolitics is the study of the relationships (political and economic) between countries and the influence of geographical factors (distance, climate, resources, etc.) on these relationships. So, geopolitics is about the power and influence of individual countries over other countries. Clearly, some countries are more powerful and influential than others. The power of a country comes from geographical features such as:

■ a large physical extent
■ a strategic global location
■ a large, well-educated population
■ a wealth of natural resources

■ a high level of economic development
■ a command of modern technology
■ strong military forces.

In this section, we will find out how these differences in political power affect three different aspects of globalisation: trade, migration and tourism.

CHECK YOUR UNDERSTANDING

Check how many of these geographical features are possessed by today's superpowers – the USA, China and Russia. Check how few are possessed by the least developed countries, such as Nepal, Somalia and Haiti.

TRADE

SKILLS   SELF DIRECTION

ACTIVITY

Choose one of the trade blocs shown in Figure 8.16. Find out:
a) when it was set up, and b) its member countries.

The most obvious sign of geopolitical relationships at work is the grouping of countries into trade blocs, such as the EU, NAFTA and ASEAN (Figure 8.16). Basically, these blocs allow free trade between member countries, while trade with outsiders is normally subject to tariffs. The terms of global trading favour these trading blocs and the superpowers. Less developed and less powerful countries are at a disadvantage. The aims of the WTO are to make global trade freer and fairer (see Figure 8.4 on page 216).

The superpowers and the TNCs are the most influential players in global trade. China is desperate to acquire all sorts of resources, especially minerals. Many African countries now export these in large quantities to China. But the geopolitical relationships mean that China is able to buy these raw

▼ Figure 8.16: Trade blocs of the world. Note that the USA may withdraw from NAFTA.

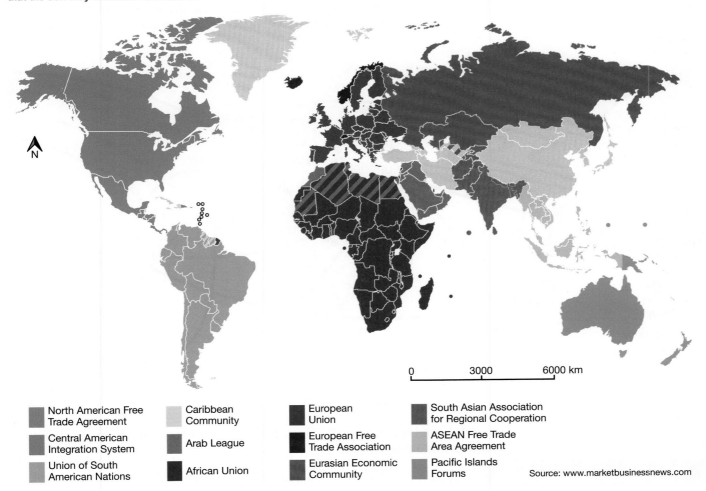

0   3000   6000 km

North American Free Trade Agreement	Caribbean Community	European Union	South Asian Association for Regional Cooperation
Central American Integration System	Arab League	European Free Trade Association	ASEAN Free Trade Area Agreement
Union of South American Nations	African Union	Eurasian Economic Community	Pacific Islands Forums

Source: www.marketbusinessnews.com

SKILLS   REASONING

ACTIVITY

Why is China so interested in the resources of Africa?

commodities for low prices. This is the strength of its bargaining power. In some instances, the commodities are obtained in exchange for China's help with major construction schemes, such as the building of new railway links or large dams.

Tensions in geopolitical relations, for example within the Middle East, are often reflected in global trade. Military equipment of all types is acquired and shipped into the region in support of the warring factions.

MIGRATION

Part 8.3 dealt with the growing volume of international migration and the reasons for it. Geopolitical relationships are important here, for a number of reasons.

- In the case of economic migrants, the migration flows are towards the more developed and powerful countries. Migrants are keen to share in the national prosperity.
- Destination countries often have the political power to control the numbers entering them. They do so by limiting the number of residential visas or work visas. Australia has a quota system, which means that it chooses migrants with the skills that are in short supply. See also the UK case study on page 234.
- When it comes to accepting refugees, in theory it is agreed that all countries should play their part in offering sanctuary. However, in practice this does not always work out. For example, compare Germany and the UK, two popular destinations for Syrian refugees. Germany has admitted over 500 000, the UK less than 20 000! The UK argues that it prefers to help Syrian refugees by setting up temporary camps close to the Syrian borders

▼ Figure 8.17: Major global migration flows (2013)

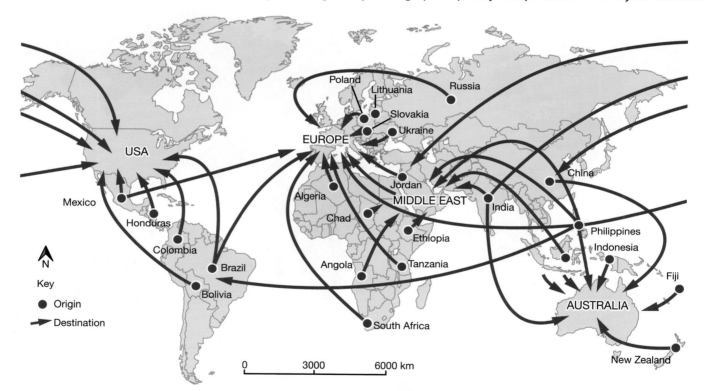

SKILLS INTERPRETATION

ACTIVITY

Look at Figure 8.17 and identify major migration hubs or focal points where flows converge.

The idea is that when the hostilities are over, these refugees will be able to return to their homes.

Figure 8.17 illustrates a number of geopolitical points.

- The 'pull' of the USA, EU and Australia as migration destinations. The attractions include: secure employment, good standard of living, a democratic society, civil liberties, etc.
- Migration flows do not appear to have been discouraged by the geopolitical factor of distance.
- The two superpowers claiming to rival the USA (China and Russia) do not attract migration to the same degree. Indeed, no flows are shown into Russia. Why is this? Might it be that both superpowers lack migrant appeal because of their repressive governments?

TOURISM

Today, the distances between countries are no longer as significant as they were 100 years ago. Package holidays have reduced the travel costs of the tourist. For some people, distance may be part of the tourist attraction. What can be more exciting than travelling to very distant and different places on the other side of the globe?

The superpowers are important players in tourism for two reasons:

■ they have large populations which enjoy annual paid holidays and have money to spend on travel and tourism
■ they themselves appeal as tourist destinations; their global reputation draws people from less powerful countries who are curious just to see what these powerful countries are like (Figure 8.18).

Geopolitical relationships are of little consequence in today's tourism, except where there is open hostility between countries. For example, for their own safety tourists are not recommended to travel to countries where there is civil unrest. Some countries in Africa and the Middle East currently fall into this category. Rather different is the example of isolationist North Korea. It does not welcome tourists, other than from China. North Koreans are not permitted to holiday abroad. Provided they have something 'special' to offer the tourist, even the poorest of countries have the opportunity to become involved in tourism. The downside for them is that the development of tourism often lies in the hands of companies based in the more developed and politically powerful countries.

SKILLS ▶ INTERPRETATION

ACTIVITY

Identify the heaviest tourist flows shown on Figure 8.18.

▼ Figure 8.18: Global tourist flows in 2010

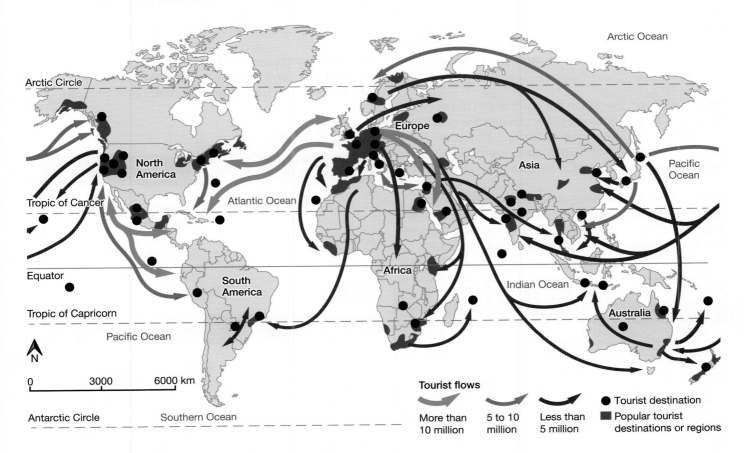

8.8 MANAGING MIGRATION

One of the many responsibilities of any government is monitoring and managing its population. A key issue is the rate of population change. Is it too fast or too slow? Whichever the answer, what needs to be done in terms of: a) natural change, and b) net migration? Rates of natural change can be influenced by promoting birth control or encouraging parents to have more children. Governments can either encourage or discourage migrants. The following case study illustrates how successive UK governments have tried to manage migration in a way to meet its changing demographic and economic needs.

CASE STUDY: THE UK'S MANAGEMENT OF IMMIGRATION

It was soon after the end of the Second World War (1945) that the UK opened its doors to immigrants, mainly from the Caribbean and from what had been the Indian Empire (India, Pakistan and Bangladesh). The UK at this time had a serious shortage of labour as so many people had been killed or badly injured in the war. The post-war reconstruction of the country to repair the massive amount of bomb damage also created a huge demand for labour. All Commonwealth (ex-colonial) citizens had free entry into the UK. By 1971, there were over 1 million Commonwealth immigrants in the UK. The government decided that was enough. So, controls were introduced to reduce the number of migrant arrivals.

Table 8.4 shows the rise in the number of UK residents born outside the UK since 1951. Despite the controls on immigration during the 1970s and 1980s, a comparison of 1971 and 1991 data clearly shows that the number of immigrants increased.

In the 1990s, the UK once again found itself short of labour. This coincided with the collapse of communism in Eastern Europe. It released huge numbers of people looking for work and a decent wage. The influx of

workers into the UK was given a boost in 2004 when the Eastern European states of the Czech Republic, Estonia, Hungary, Latvia, Lithuania, Poland, Slovakia and Slovenia joined the EU. Figure 8.19 shows the push-pull factors. In many cases, these economic migrants intended to stay only until they felt that they had made enough money to take home. Few intended to stay permanently. Figure 8.20 shows that well over half the migrants came from Poland, the largest of the new member states. The vast majority of migrants were young and single, with over 80 per cent aged 18 to 34.

With net immigration running at 300 000 persons per year, the control of immigration was one of the key issues in the UK's Referendum about continuing

SKILLS ▶ DECISION MAKING

ACTIVITY

In class, discuss the issue of whether or not governments should control immigration.

SKILLS ▶ PROBLEM SOLVING (GRAPHICAL)

ACTIVITY

Plot the data in Table 8.4 on a graph and summarise what the completed graph shows.

CENSUS	FOREIGN-BORN POPULATION (MILLION)	PERCENTAGE INCREASE SINCE PREVIOUS CENSUS	PERCENTAGE OF TOTAL POPULATION
1951	2.1		4.2
1961	2.6	21.5	4.9
1971	3.2	24.0	5.8
1981	3.4	7.5	6.2
1991	3.8	11.8	6.7
2001	4.9	27.7	8.3
2011	8.0	63.0	12.7

◀ Table 8.4: The growth of the foreign-born population of the UK (1951-2011)

membership of the European Union held in 2016. The majority vote was that the UK should leave the EU and be more in charge of its own affairs. At the moment, it is only able to control the number of immigrants coming from outside the EU. In the future, it looks as if people from EU countries will no longer have the right to freely enter the UK. In theory, the UK government will then be better able to monitor immigration.

▲ Figure 8.19: Eastern European workers – push-pull factors

▶ Figure 8.20: The country of origin of the UK's Eastern European economic migrants (2010)

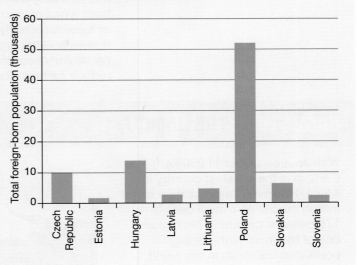

This case study clearly shows that the UK's policy on immigration has fluctuated over time. The policy has been largely determined by the labour situation. Today, there is mounting public concern in the UK about the volume of immigration (compare the 2001 and 2011 data in Table 8.4) and the need to control it. Membership of the EU does not allow the UK to control the influx of economic migrants from other member countries. That situation is expected to change, however, with the UK's exit from the EU.

8.9 MAKING TOURISM MORE SUSTAINABLE

Although there have been economic benefits, the rise of tourism has also put great pressure on popular tourist destinations, as illustrated by the Costa Blanca case study (page 229). This has led to a general move to make this global industry more sustainable and so minimise its negative impacts. This will ensure that future generations can enjoy the same amenities – clean seas and beaches, fine natural scenery and local cultures and their heritage.

An important part of making tourism more sustainable has been the growth of **ecotourism**. Ecotourism is the opposite of mass tourism. It often involves small numbers of tourists visiting locations that are often relatively inaccessible (Figure 8.21).

The main features of ecotourism are:

- it is based in locations that are in some way special because of their scenery, wildlife, remoteness or culture
- it aims to educate people and increase their understanding and appreciation of nature and local cultures
- it tries to minimise the consumption of non-renewable resources and environmental damage
- it is locally-oriented – controlled by local people, employing local people and using local produce

▲ Figure 8.21: An environmentally-friendly tree-canopy walkway

- its profits stay in the local community
- it is sustainable and it contributes to the conservation of biodiversity and culture.

It is definitely a much 'greener' form of tourism than mass tourism. However, both types of tourism are challenged by the same issue, namely the burning of fossil fuels to fly and drive the often huge distances between the tourists' homes and the growing global network of tourist destinations. The issue is particularly sensitive with concern about climate change and the need to reduce carbon dioxide emissions.

CASE STUDY: THE GAMBIA

With an area of just 11 000 km², The Gambia is the smallest country in Africa. However, its population of 1.8 million means an overall density of 176 persons per km². This makes it one of the more high-density populated countries in the world.

The Gambia has few natural resources and three-quarters of the population rely on agriculture. This accounts for 30 per cent of GDP, but it employs 70 per cent of the labour force. Equal second as 20 per cent contributors to GDP are remittances (money sent home by Gambians working abroad) and tourism. The Gambia is a popular 'sand, sea and sun' destination for around 100 000 tourists each year. Most come from Europe which is only a six-hour flight away.

▲ Figure 8.22: An eco-lodge beside the Gambia River

While tourism brings in foreign currency and provides employment, it is difficult to make it more sustainable:

- tourism is in the hands of European package tour operators, so most of the profits 'leak' out of the country
- there are few untouched areas that might be converted, given special protection and designated as natural reserves

- feeding the tourists provides the stimulus for some commercial farming, but many of the goods and services needed to support tourists during their stay have to be imported
- tourism is concentrated on the coast; some efforts are being made to attract more visitors upriver by building eco-lodges and camps that at least allow the visitors a more truly African experience (Figure 8.22).

One simple and achievable objective should be to minimise the negative environmental and social impacts of tourism. For example, by ensuring that hotels do not pollute the waters of the river and coast; and by respecting the cultural traditions and codes of conduct of a largely Islamic population.

SKILLS　REASONING

ACTIVITY

What makes Figure 8.22 an example of eco-tourism?

CASE STUDY: BHUTAN – ECOTOURISM ON A NATIONAL SCALE

There are a growing number of ecotourism ventures around the world pioneering this mode of tourism and demonstrating its benefits. Almost all of these projects are the outcome of private enterprise. Whilst governments often give verbal support, little action to promote ecotourism has been taken so far. One notable exception is the remote Himalayan Kingdom of Bhutan.

Bhutan's tourist attractions are spectacular mountain scenery and a rich Buddhist cultural heritage of ancient temples and shrines. These sorts of attractions are never likely to appeal to mass tourists. However, to alternative tourists, such as trekkers, birdwatchers and those with an interest in cultural history, Bhutan has a fascination (Figure 8.23).

Tourism began in Bhutan in 1974 when the King realised that the new hotels built to accommodate guests at his coronation could be used for tourism. Tourism, in turn, would generate foreign exchange and provide the means for the country's economic development. The decision to become involved in global tourism was a brave one. Bhutan had only just opened its doors to foreigners after 300 years of isolation. Initially, the number of foreign visits was limited to 2500 a year. That limit has since been more than doubled. All tourists must be part of an escorted group to specified locations. Tourists are required to pay a surcharge that currently stands at US$250 a day per person, with a surcharge of US$40 per person for those travelling alone or as one of a couple. All tours must be organised by known, vetted companies. All developments, such as hotels, must use traditional architectural designs. The emphasis is on conservation of the natural environment and culture.

The country's attitude to tourism is ambivalent. It is keen to get the economic benefits, while still viewing tourism as undesirable. The government's tourist strategy tightly controls the volume and potential impacts of tourism.

SKILLS SELF DIRECTION

ACTIVITY

Look at an atlas map of the Indian sub-continent and locate the Kingdom of Bhutan.

SKILLS DECISION MAKING

ACTIVITY

As far as ecotourism is concerned, 'small is beautiful'. So, do you agree that there is likely to be a tension between this fact and any attempt to expand ecotourism at a national scale?

◀ Figure 8.23: Bhutan is an alternative tourist destination.

CASE STUDY: CYPRUS

The island of Cyprus has been a major tourist destination for Europeans for a long time. Tourists are drawn by its climate, heritage and coastal attractions. Over 2 million tourists arrive each year; July is the peak month. In this case study, attention will focus on the Greek part of this politically divided island.

The Cyprus Sustainable Tourism Initiative (CSTI) was launched in 2006. Its purpose is to encourage a more sustainable form of tourism through:

■ preserving, conserving and protecting the environment
■ a better, more efficient use of natural resources
■ reducing the carbon footprint of tourism
■ improving the conditions of those communities disadvantaged by tourism.

Unfortunately, by the time this initiative was launched much irreparable damage had already been done to the environment in Cyprus, particularly around the coast. Many poor communities had been displaced from their traditional landholdings by the building of hotels and roads.

Some current projects include:

■ 'greening' the beaches – maintaining the overall cleanliness of beaches by organising beach-cleaning events; ensuring that in-shore waters are not polluted (Figure 8.24)
■ encouraging rural tourism away from the coast – developing the skills of women in various tourist activities, such as providing accommodation, producing arts and crafts and preparing local foods

▲ Figure 8.24: A clean-up operation on a Cyprus beach

■ managing water, energy and waste – particularly reducing levels of water consumption (for example, in swimming pools), relying more on renewable energy (solar), and reducing the use of plastic and therefore the amount of plastic waste.

These are encouraging early steps in the right direction. But there is little hope of reducing the **carbon footprint** of tourism, since most of the tourists here, and in other tourist hotspots, arrive by air. And once here, they like to tour the island by coach or car.

CHECK YOUR UNDERSTANDING

Check that you understand what is meant by the term carbon footprint.

These three case studies are encouraging in that they show an increasing awareness of the need for a more sustainable form of tourism in different parts of the world. Sadly, tourism has done much damage in the long-established and popular tourist destination countries, such as Thailand, Cyprus and Spain. Here, much needs to be done, but it is not too late to start pressing for a more sustainable form of tourism. Interestingly, it is one of the poorest countries in the world, Bhutan, which can claim to have one of the most sustainable tourist industries in the world. Bhutan offers a beacon of hope, particularly to those developing and emerging countries which look to tourism playing a part in their economic development.

CHAPTER QUESTIONS

END OF CHAPTER CHECKOUT

INTEGRATED SKILLS
During your work on the content of this chapter, you are expected to have practised in class the following:

- using and interpreting line graphs and bar charts showing changes in the global economy over the last 50 years
- using maps to identify patterns of migration
- interpreting photographs and newspaper articles
- using and interpreting graphs showing rates of population movement over the last 50 years
- using and interpreting socio-economic data.

SHORT RESPONSE

1　Name **two** major players in the global economy.　[2]

2　State **two** characteristics of transnational companies (TNCs).　[2]

3　State **one** example of voluntary migration.　[1]

4　Identify **one** characteristic of ecotourism.　[1]

LONGER RESPONSE

5　Identify **three** geographical features that contribute to the geopolitical power of a country.　[6]

6　Examine reasons for the rise in retirement migration.　[6]

7　Examine the possible impacts of refugees on the country receiving them.　[12]

8　Assess possible ways of making tourism more sustainable.　[12]

EXAM-STYLE PRACTICE

1　**a)** Identify **one** reason for refugee migration.　[1]
　A　Old age
　B　Poor housing
　C　Need for a job
　D　Persecution

　b) State **one** type of area that is attracting retirement migration.　[1]

2　**a)** Name **one** trade bloc.　[1]

　b) Suggest why it has been formed.　[2]

　c) Identify **two** benefits to a country having a TNC within its borders.　[2]

3　Explain why a government might wish to control immigration.　[6]

4　**a)** Study Figure 8.11 (on page 226). Calculate Europe's percentage of world tourist arrivals.　[2]

　b) Suggest **two** reasons for Europe's high percentage figure.　[2]

5　Explain the advances in technology that have helped globalisation.　[6]

6　Evaluate the view that the economic benefits of tourism outweigh environmental costs.　[12]

[Exam-style practice total 35 marks]

9 DEVELOPMENT AND HUMAN WELFARE

This chapter is about two important aspects of the modern world – development and the welfare of people. Development is certainly bringing change to virtually all parts of the world. However, the level of development varies greatly between countries and within countries. So, too, do its different impacts, especially on human welfare. There are significant differences between rich and poor, and between successful and not-so-successful countries and regions. The unevenness is reflected not just in the quality of life, but also in the characteristics of populations. How is this unevenness to be reduced? Opinions are divided on possible strategies and approaches.

9.1 DEFINING DEVELOPMENT AND HUMAN WELFARE

Development and human welfare are closely interwoven (Figure 9.1). This is best explained if we first define both terms.

▶ **Figure 9.1 One happy family – an advert for development and human welfare?**

DEVELOPMENT

Development is a process of change that affects countries and their peoples. Just as people grow up, so do countries. Both processes are about maturing and becoming stronger and more independent.

SKILLS REASONING

ACTIVITY

Is globalisation helping development? If so, how?

Development means making progress in a number of different fields. We can call these fields the 'strands' of development. By far the most important strand is an economic one – economic development. This provides the power that drives progress in all the other development strands. Economic development comes from the exploitation of resources – minerals, energy, climate and soils. It also requires capital, technology and, above all else, enterprising people and good government. The other strands of development fall into five groups. Table 9.1 gives more details.

▼ **Figure 9.2: Factors generating economic development**

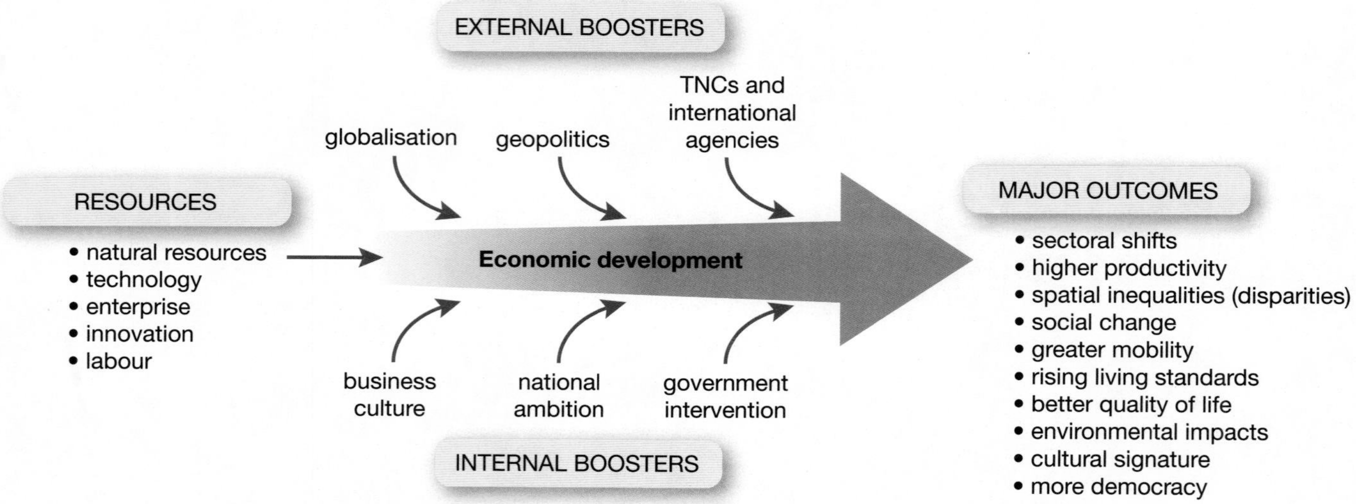

Advances in all or most of these strands mean that a country is developing. This could include a rising standard of living, greater life expectancy, more welfare services, democratic government and more respect for the environment.

EXTERNAL BOOSTERS

globalisation   geopolitics   TNCs and international agencies

RESOURCES
- natural resources
- technology
- enterprise
- innovation
- labour

Economic development

business culture   national ambition   government intervention

INTERNAL BOOSTERS

MAJOR OUTCOMES
- sectoral shifts
- higher productivity
- spatial inequalities (disparities)
- social change
- greater mobility
- rising living standards
- better quality of life
- environmental impacts
- cultural signature
- more democracy

▼ Table 9.1: Some strands of development

GROUP	DEVELOPMENT STRAND
Economic	Employment: security and levels of pay Standard of living: raising the minimum Productivity: efficient use of capital and labour
Demographic	Life expectancy: rising with better healthcare, hygiene and diet Birth control: right to choose family size Mobility: freedom to migrate
Social	Welfare: access to services Equal opportunities Quality of life
Cultural	Education: compulsory education for all Heritage: respect and conserve Ethnicity: mutual respect
Political	Right to vote Democratic government: regular and fair elections Freedom of speech
Environmental	Pollution: effective controls Conservation: biodiversity and non-renewable resources Ecological footprint: minimising this

SKILLS ▶ REASONING

ACTIVITY

Suggest **two** reasons why economic development provides the power that drives development as a whole.

It is widely agreed that economic development lies at the heart of development. Figure 9.2 shows how economic development works. There are three main forces involved.

■ Resources – these get the process moving in the first place. There are natural resources (such as soils, climate and minerals) and human resources (such as enterprising business people, capital, labour and technology). Human resources exploit natural resources and, in doing so, provide the drive behind economic development.

■ Internal boosters – these come from within the individual country and include government intervention, national ambition, the growth of a business culture, etc.

■ External boosters – these come from outside the individual country and include the growth of the global economy that creates opportunities for countries to prosper. Other boosters include key players in the global economy such as the transnational corporations (TNCs) and various international agencies and organisations. For more on the global economy, see Part 8.1, page 214.

Figure 9.2 shows the outcomes of economic development. The list shows some of the signs, and possible measures, of development (see Part 9.2).

HUMAN WELFARE

Like development, human welfare is a multi-strand condition. It is defined simply as the general condition of a population or society. It is very difficult to define it more precisely than that. Most people prefer to use the term **quality of life**, which probably covers most human welfare issues. Other terms used in studies of human welfare include well-being and standard of living.

In this chapter we will use the term quality of life. The problem with this term is that it is not only about material things, such as income and housing. It is also to do with how people feel about their lives. Do they feel secure and safe? Do they enjoy good health? Are they happy? These sorts of questions mean that quality of life has a psychological aspect (Figure 9.3).

This psychological aspect is only one of four components making up quality of life. All four are interrelated. For example, look at the 'Economic' box in Figure 9.3. Whether or not we feel secure will hinge on whether we have a secure job and a regular income. These, in turn, will affect our level of **affluence** and our standard of living. In the 'Physical' box, **diet** and housing can affect our state of health. In the 'Social' box, a good education will play a considerable part in securing a secure job and enjoying a good standard of living.

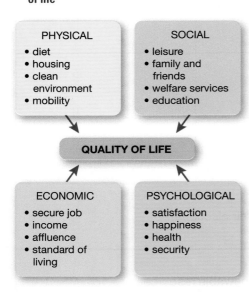

▼ Figure 9.3: Some components of quality of life

An improving quality of life is one of the outcomes of economic development (see Figure 9.2). The link between the two is a simple one. Economic development involves creating all sorts of wealth. This then starts a cycle of wealth (Figure 9.4). Part of the wealth goes to the government in the form of taxes. If there is good governance, this money will be spent on goods such as roads, defence, education and **healthcare**.

▶ Figure 9.4: The cycle of wealth

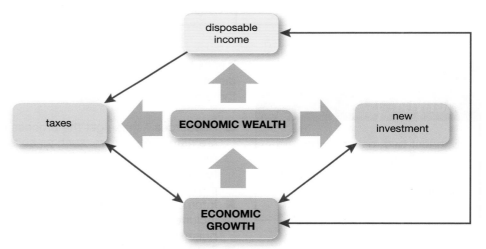

DID YOU KNOW?

As with development, it is the economic aspect that is the most critical aspect of quality of life. Yes, we do live in a material world, and yes, money talks…

SKILLS INTERPRETATION

ACTIVITY

What is the link between disposable income and taxes, shown in Figure 9.4?

Another part of wealth comes in the form of work and wages. This gives people a disposable income that they can spend on a range of goods and services, such as shops, restaurants, holidays and hairdressers. Workers and their families should benefit from the basic services funded by the government, such as schools and doctors. In these two ways, the quality of life improves for people. Workers also pay taxes which help to pay for those basic services.

A third part of that created wealth may be reinvested by businesses to expand their operations (Figure 9.4). This starts a new cycle of wealth leading to further improvements in the general quality of life. Unfortunately, increasing wealth and improving quality of life are not equally shared by everyone. Economic development often results in a widening gap between the rich and the poor (see Part 9.4). The benefits of economic development are not shared

9.2 CONTRIBUTORS TO DEVELOPMENT AND HUMAN WELFARE

Let us look more closely at some of the factors shown in Figure 9.2.

NATURAL RESOURCES

Resources must be at the top of the list when it comes identifying things that contribute most to development. With the right resources, there should be no stopping a country's development. Resources are of two kinds – natural and human.

▶ Figure 9.5: Sources of natural resources

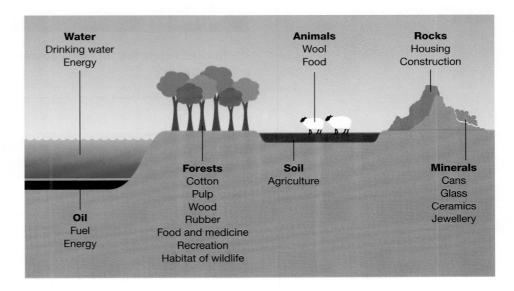

Natural resources are those provided by the physical environment (Figure 9.5). A country needs those resources to deliver security in three key areas – water, food and energy. Ideally, a country should meet all of its demand for these three resources from within its borders. But there are few countries in this enviable situation. Most countries have to import some food or energy. Some countries even have to import water. The more imports, the greater the insecurity. Why? Because importing countries are vulnerable to several risks. For example, supplies might be cut off by wars or for political reasons. Prices might be deliberately raised, as a form of blackmail, to put pressure on the importing country for some ulterior motive.

▼ Figure 9.6: Education – the key to improving human resources

HUMAN RESOURCES

Many people believe that technology is the most important of the human resources. Technology can help increase water, food and energy security by finding new ways of using domestic resources more efficiently or by discovering alternatives. Any innovation that reduces imports will be good for security.

SKILLS ▷ REASONING

ACTIVITY

The quality of a country's labour force is also significant. Key labour qualities include skills, enterprise and commitment.

OTHER CONTRIBUTORY FACTORS

- Government – development probably flourishes best if the government is democratically elected and encourages enterprise and private investment. Whether the government or political regime is capitalist or socialist might make some difference. The world's two largest economies, the USA and China, each represent one of those regimes. Although at the top of the global rankings, they are very different in terms of per capita GDP and possibly quality of life. Political stability is important in both situations.
- Equality – where there is equality, a country's progress will not be held back by tensions, divisions and conflicts. Equality applies to different aspects of a society, from opportunities and gender to ethnicity and class.

ACTIVITY

To what extent is your home country secure in terms of water, food and energy?

9.3 MEASURING DEVELOPMENT AND QUALITY OF LIFE

DEVELOPMENT

Can we measure development in a way that allows us to compare countries in terms of their level of development? Given that development is a multi-strand process, measuring it will not be easy. Do we have to measure all the strands separately, or are some strands more important and better indicators than others? If so, which are they?

▶ **Figure 9.7: The strands of the development cable**

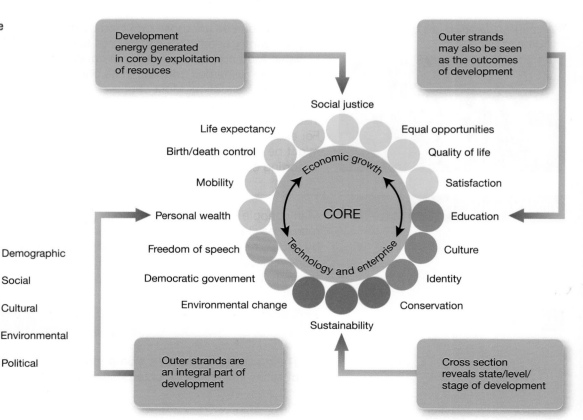

Development energy generated in core by exploitation of resouces

Outer strands may also be seen as the outcomes of development

Social justice

Life expectancy

Equal opportunities

Birth/death control

Quality of life

Mobility

Satisfaction

Personal wealth

Economic growth

CORE

Education

Freedom of speech

Technology and enterprise

Culture

Democratic govenment

Identity

Environmental change

Conservation

Sustainability

Key

Demographic

Social

Cultural

Environmental

Political

Outer strands are an integral part of development

Cross section reveals state/level/stage of development

CHECK YOUR UNDERSTANDING
Check that you understand the difference between GDP and GNI.

SKILLS INTERPRETATION

ACTIVITY

What conclusions do you draw from Table 9.2?

Economic development drives development as a whole, so our search for measures should start with this. There are two indicators widely used to assess the strength of a country's economy.

- Gross domestic product (GDP) – the total value of a country's economic production over the course of a year.
- Gross national income (GNI) – this differs from GDP in that it includes the total value of a country's economic production plus net income received from abroad. It, too, is calculated for a year. It used to be called gross national product (GNP).

However, these two indicators do not take into account that countries vary enormously in size. As Table 9.2 shows, a large country is likely to have a large GDP. However, if we divide the GDP or GNI values by the number of people in a country, we arrive at two measures that allow us to compare countries in a sound way. These indicators are known as per capita GDP (or GDP per head of population) and per capita GNI (or GNI per head of population). Table 9.2 illustrates that two large countries can show very different per capita GDP values. The same also applies to small countries.

COUNTRY	AREA (MILLION KM²)	POPULATION (MILLION)	GDP ($US TRILLION)	PER CAPITA GDP ($US)
USA	9.8	325	18.5	55 836.8
China	9.6	1357	11.4	7924.7
Haiti	0.28	10	0.1	819.9
UK	0.24	65	2.8	43 734.0

▲ Table 9.2: Some international comparisons of per capita GDP (2015)

As the economy of a country develops, not only do GDP and GNI become larger, but how that economic wealth is produced changes. In the early stages of economic development, it is the primary sector that generates the most growth. Agriculture, fishing, forestry and mining are the mainstays of this economy. Gradually, the secondary sector becomes the main generator of economic growth. Raw materials are manufactured into goods that have a higher value than food, fish or minerals. As personal wealth increases, the tertiary or service sector takes over as the most important part of the economy. In most developed countries, a new sector is now appearing. This is the quaternary sector, which is based on information and communications technology (ICT) and research and development (R & D). Although growing, this sector is still greatly overshadowed by the tertiary sector as a source of economic growth.

DID YOU KNOW?
For more information about sector shifts, see Part 4.1.

From this sector shift, we can assess a country's level of economic development. We do this by measuring the relative importance of the three or four sectors. Figure 4.14 (on page 108) illustrates this. Ethiopia is a country that has made little progress in terms of economic development. Note that the primary sector is the largest. China is a country experiencing rapid economic development. Although the tertiary sector appears larger than the secondary sector in China, it is the secondary sector that manufactures the goods being sold across the world. These sales are the major source of China's economic growth. The UK is more economically advanced than China – the economy is dominated by the tertiary sector. Look how small the primary sector is.

▲ Figure 9.8: Electricity – the power behind much economic development

Calculating the relative importance of the economic sectors can be based on two different measures. It is based either on the number of people employed in each sector, or how much each sector contributes to the economy (see Figure 4.14).

ACTIVITY

Explain why electricity is so important to development.

Other possible indicators of development are listed here.

■ Energy consumption – the greater the economic development of a country, the greater its consumption of energy for manufacturing and transport. Energy consumption is also increased by the use of electricity in the home and to power many services (such as air conditioning in shops and offices, street lighting and telecommunications).

■ Population rates – with economic development, birth rates fall due to increased birth control. Death and **infant mortality** rates also fall as a result of advances in medicine and healthcare, and of people living in better housing and having a better diet. As a consequence of this 'death control', life expectancy increases.

QUALITY OF LIFE

Since quality of life, like development, is made up of many strands, we come up against the same questions as in Part 9.2. How do we measure it? What indicators should we use? Figure 9.9 takes three strands – housing, education and health. For each, it shows six indicators or possible measures. The three world maps (Figures 9.10, 9.11 and 9.12) are each based on one measure for each of those strands – dwellings with access to safe drinking water, literacy rate and daily food consumption.

▼ Figure 9.9: Some quality of life indicators

Housing

- Dwelling floor-space per capita
- Percentage of dwellings with running water
- Percentage of dwellings with electricity
- Percentage of dwellings with an indoor toilet
- Percentage of dwellings owner-occupied
- Percentage of income spent on house

Education

- Percentage of GNP (or GDP) spent on education
- Average number of years in full-time education
- Full-time students per 1000 people
- Literacy rate
- Full-time teachers per 1000 people
- Percentage of school leavers going on to higher education

Health

- Infant mortality rate
- Life expectancy
- Percentage of GNP (or GDP) spent on health
- Doctors per 1000 people
- Hospital beds per 1000 people
- Daily intake of calories

▼ Figure 9.10: Access to safe water

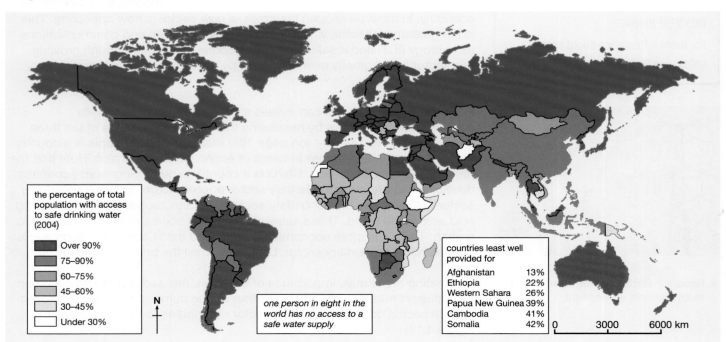

the percentage of total population with access to safe drinking water (2004)

- Over 90%
- 75–90%
- 60–75%
- 45–60%
- 30–45%
- Under 30%

one person in eight in the world has no access to a safe water supply

countries least well provided for

Afghanistan	13%
Ethiopia	22%
Western Sahara	26%
Papua New Guinea	39%
Cambodia	41%
Somalia	42%

0   3000   6000 km

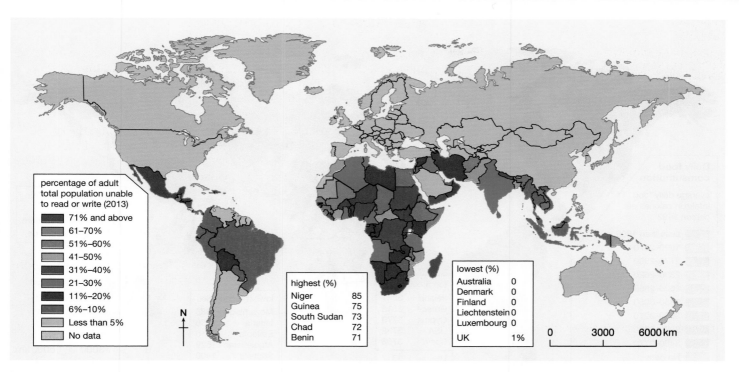

percentage of adult total population unable to read or write (2013)

- 71% and above
- 61–70%
- 51%–60%
- 41–50%
- 31%–40%
- 21–30%
- 11%–20%
- 6%–10%
- Less than 5%
- No data

highest (%)

Niger	85
Guinea	75
South Sudan	73
Chad	72
Benin	71

lowest (%)

Australia	0
Denmark	0
Finland	0
Liechtenstein	0
Luxembourg	0
UK	1%

0   3000   6000 km

▲ **Figure 9.11: The global distribution of illiteracy**

SKILLS INTERPRETATION

ACTIVITY

Look back at Figure 9.3 (page 244) and identify any other components of quality of life that you think can be measured accurately.

■ Housing – shelter is a basic human need. Housing conditions will directly affect quality of life, particularly if all the basic services (clean or safe water, sewage disposal and electricity) are available and the density of occupation (the number of people per unit of housing space) is relatively low.

■ Literacy – education is thought to be the key to a better quality of life. It opens the door to regular employment. This means economic security, as well as disposable income to buy those things that improve the quality of life. The percentage of a population able to read and write is a good indicator of the general level of education in a country. Figure 9.11 looks at the situation in the opposite way. It shows the global distribution of illiteracy, that is the number of people who are unable to read or write.

■ Diet – an adequate diet is important for health. Poor health will result from not eating properly. This can have a strong negative impact on how people feel and their general outlook on life. Poor health can also mean time off work which can damage economic security.

A daily food consumption below 2000 calories often means malnutrition, Figure 9.12 shows that at least nine African countries whose populations fall into this category, together with Peru in South America and Afghanistan, Nepal and Mongolia in Asia.

Are these three indicators (access to clean water, literacy rates and daily food consumption) of equal importance? Do they all show the same global distribution patterns? Can we come up with a single measure that provides a reliable indicator for a sound comparison of regions and countries?

SKILLS INTERPRETATION

ACTIVITY

In which parts of the world is malnutrition a serious problem? Look at Figure 9.12 on page 250 for the answer.

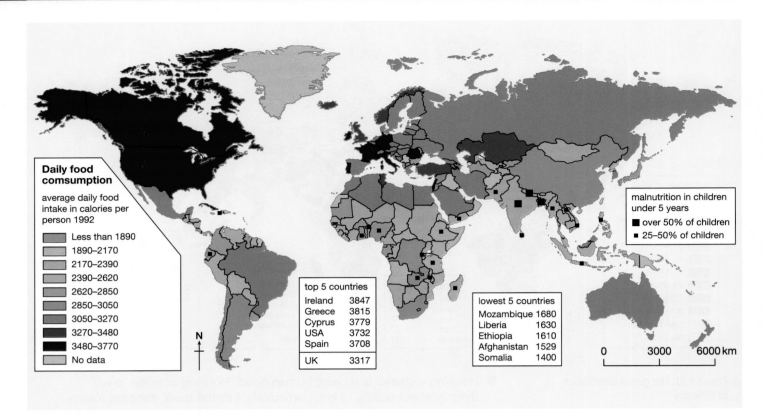

Daily food consumption

average daily food intake in calories per person 1992

	Less than 1890
	1890–2170
	2170–2390
	2390–2620
	2620–2850
	2850–3050
	3050–3270
	3270–3480
	3480–3770
	No data

malnutrition in children under 5 years
- over 50% of children
- 25–50% of children

top 5 countries	
Ireland	3847
Greece	3815
Cyprus	3779
USA	3732
Spain	3708
UK	3317

lowest 5 countries	
Mozambique	1680
Liberia	1630
Ethiopia	1610
Afghanistan	1529
Somalia	1400

0   3000   6000 km

N

▲ Figure 9.12: Daily food consumption

SKILLS ▶ INTERPRETATION

ACTIVITY

Look at Figure 9.13:

a) Can you identify and name the scattered countries in Asia shown with high HDI values?
b) Which parts of the world have 'medium' HDI values?

SKILLS ▶ REASONING

ACTIVITY

Suggest reasons for the high index values shown on Figure 9.14.

Human Development Index (HDI) is the most widely used aggregate measure of the quality of life. It is also used as a measure of the level of development. The HDI, like the PQLI, also only takes into account three variables. With the HDI, these three are per capita income, literacy and life expectancy. Therefore, the HDI assesses three rather than two different strands: one economic, one education and one health. The calculation of the HDI is a little complicated, but it assumes that the three variables are of equal importance. The HDI is the average of the scores achieved by a country in those three strands. HDI scores range from 0 to 1. The higher the HDI, the higher the level of development and the better the quality of life.

Figure 9.13 shows how HDI values vary at a global scale. HDI values are high in North America, southern South America, Europe and Australasia. At the other end of the HDI scale, Africa is the main focus of low and unsatisfactory values.

Finally, we will look at two other measures which can apply to both development and quality of life.

- Gini index – this measure of inequality is used to analyse the distribution of wealth or income among the citizens of a country. A low index value indicates distribution that is close to equal (Figure 9.14). A high value indicates inequality between rich and poor.
- Index of corruption – this scores each country on how corrupt their government is seen to be by researchers working for Transparency International. Figure 9.15 on page 252 shows the results of the 2014 survey. The scaling runs from 'clean' (high index value) to 'most corrupt' (low index value). Asia, Africa and much of Latin America are shown as having corrupt governments.

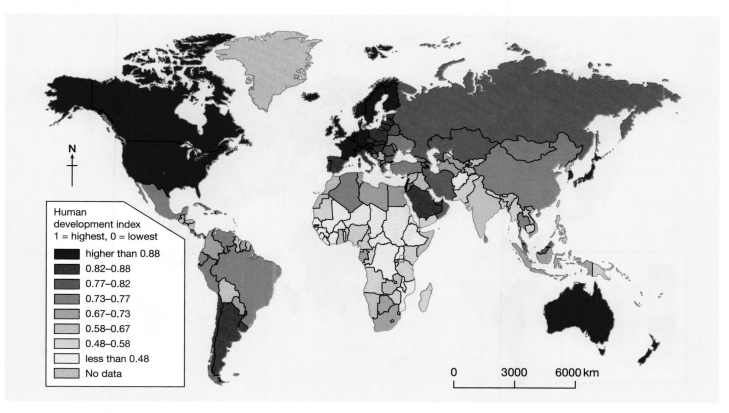

▲ Figure 9.13: The Human Development Index

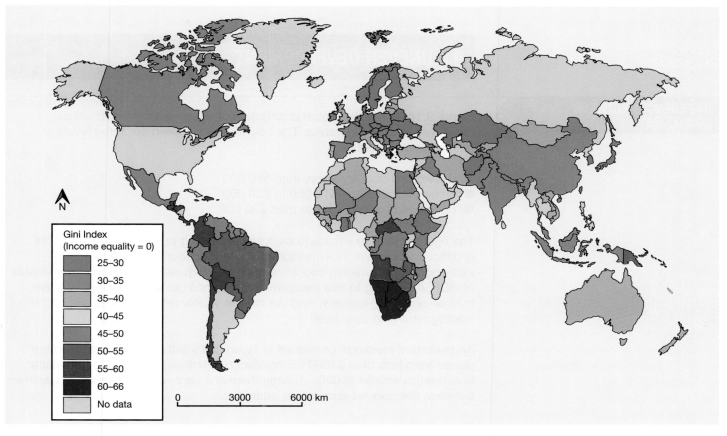

▲ Figure 9.14: Income equality

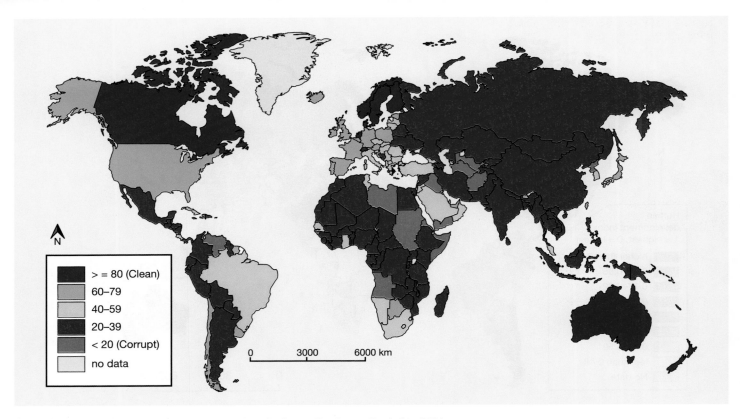

> = 80 (Clean)
60–79
40–59
20–39
< 20 (Corrupt)
no data

0   3000   6000 km

▲ Figure 9.15: A global view of corruption based on the Corruption Perception Index, 2014

9.4 UNEVEN DEVELOPMENT

DIFFERENCES BETWEEN COUNTRIES

Figure 9.16 shows the global distribution of economic development as measured in GDP per capita. The boundaries between our three levels of development would be:

- developing country – less than $10 000
- emerging country – $10 000 to $30 000
- developed country – more than $30 000.

The figures relate to what is known as purchasing power parity (PPP). This is difficult to explain. The measure takes into account different currency exchange rates between countries. Using PPP gives a more accurate measure of GDP. According to this measure, India comes out as a developing rather than an emerging country. And yet India is widely recognised as one of the leading emerging countries!

An important message contained in Figure 9.16 is the extreme range in the values from less than $1000 (eight African countries) to over $60 000 (Qatar is at the top with $132 000). Clearly there is a very wide development gap here between the poorest and richest countries.

Figure 9.13 (on page 251), based on the HDI, shows a similar pattern of unevenness with respect to human welfare.

▶ Figure 9.16: Global view of GDP per capita (PPP) (2015)

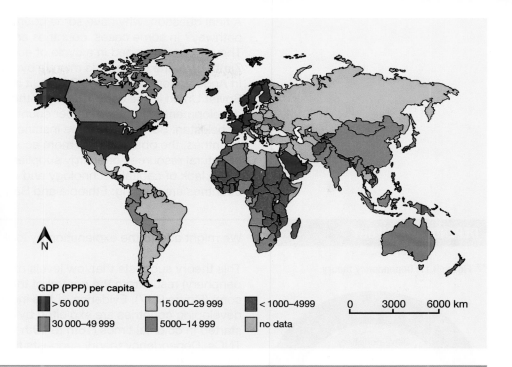

GDP (PPP) per capita

■ > 50 000	■ 15 000–29 999	■ < 1000–4999
■ 30 000–49 999	■ 5000–14 999	■ no data

0　　3000　　6000 km

SKILLS INTERPRETATION

ACTIVITY

Look at Figure 9.16. Do the leading emerging countries – Brazil, Russia, India and China – show the same GDP per capita values?

DEVELOPMENT PATHWAY

This unevenness of development across the world allows us to think of development as a pathway – running from least developed to the most developed. As they develop, countries move along that pathway but at very different speeds. At any one time, countries will be strung out along it, perhaps clustering at points. The threefold classification of countries is an encouraging one. It suggests that the world is no longer divided into two groups: rich and poor. There is a group of once-poor countries (now emerging) that are experiencing significant and relatively fast economic growth.

So, what are the factors that explain this unevenness? Why are countries now strung out along the development pathway? Presumably they all started at the same end. We have already identified some reasons in Figure 9.2 (see page 242). The UK was one of the first countries to make its way along the development pathway. Important steps were taken in the 16th, 17th and 18th centuries which led to the growth of the British Empire. The Empire gave the UK access to resources in many different parts of the world. In the late 18th century came the Industrial Revolution. This was largely based on the UK exploiting its own resources (coal and iron ore). But human resources, such as enterprise, innovation and labour, were equally important, if not more important. So part of the reason for the UK's status as a developed country was that it, along with other western European countries and the USA, made an early start along the pathway.

ACTIVITY

Write your own definition of 'development pathway'.

SKILLS REASONING

ACTIVITY

Suggest reasons why some countries have moved faster along the development pathway than others.

The next question is: why are the emerging countries now making good progress along the development pathway? A major force in driving economic development in these countries has been the global shift in manufacturing. Manufacturing companies originally located in the UK and other developed countries have been attracted to emerging countries by a number of pull factors. By far the most important of these has been the huge difference in labour costs. Labour costs in the emergent economies are a small fraction of what they are in developed countries. Manufacturers will locate where they will make the most profit.

A final question: why have some countries scarcely started along the pathway? In some cases, countries are held back by some sort of obstacle. They become trapped in a cycle of economic stagnation. In North Korea and Zimbabwe, the obstacle is misrule by undemocratic or corrupt government. In Angola, Somalia and the Sudan, it is civil war. The break-up of the former Soviet Union in the early 1990s put the brake, although only temporarily, on development in former member countries such as Romania, the Ukraine and Uzbekistan. So, in these three instances, the obstacles are political. In other countries, the obstacles are more economic. The obstacles include: a lack of natural resources or energy supplies, a lack of educated or trained labour, and a lack of capital, technology and enterprise. The obstacles are what these countries are missing; Ethiopia and Bangladesh would fall in this category.

DEPENDENCY THEORY

We might add to the explanation by looking at the dependency theory.

▼ Figure 9.17: Dependency theory

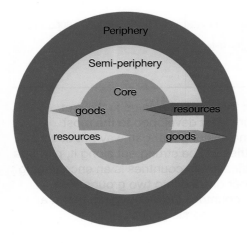

This theory suggests that low levels of development in poorer countries (the periphery) result from the control of the global economy by rich countries (the core) (Figure 9.17). Evidence of this includes the way in which the resources of developing countries are exploited by emerging and developed countries. This started in colonial times (16th to 19th centuries). It is continued today by the TNCs. Dependency theory suggests that the uneven pattern of development has been encouraged by:

- developed countries interfering with the internal politics of developing countries
- unfair trade, where developing countries sell resources cheaply but buy expensive products
- the selling of non-essential products, such as soft drinks, to developing countries
- bilateral aid that comes as part of wider agreements so that the developed country gets something in return, such as access to cheap resources
- developing countries becoming deep in debt as a result of taking out large loans from the developed world; the trouble with loans is that they have to be repaid and interest is charged while they are being repaid.

CHECK YOUR UNDERSTANDING

Can you name one developing country that was not colonised by Europeans?

But the blame for the poverty of developing countries cannot all be placed on colonialism and capitalist TNCs. Some poor countries were never colonised by Europeans; other poor countries blame socialism.

UNEVENNESS WITHIN COUNTRIES

Development levels vary between countries, and also within countries. Although **development indicators** allow comparisons of development levels between countries, it is very important to remember that these are average figures. National data does not show the variations that occur within countries and between regions. Rarely is economic wealth shared equally between all the regions of a single country. Very often, economic development and wealth are concentrated in just one favoured region.

This unevenness is especially true of large countries such as the USA. Figure 9.19 shows differences between the individual states. High HDI values in the north-east contrast with relatively low values in the south. But even much smaller developed countries, such as Italy, show some differences, namely between the north and south (see Part 9.5, and Figure 9.18). Similar unevenness of development is found in emerging countries, such as India (Figure 9.20).

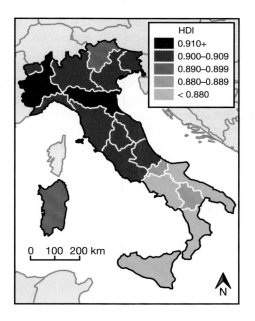

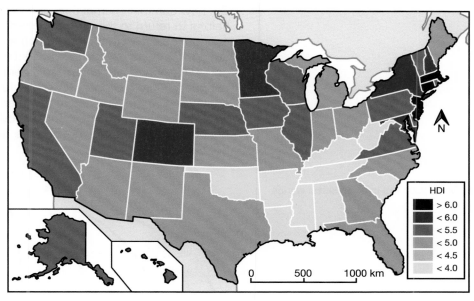

▲ Figure 9.18: Uneven development in Italy

▲ Figure 9.19: Uneven development in the United States

► Figure 9.20: Uneven development in India

SKILLS REASONING

ACTIVITY

Suggest reasons for the low HDI values in the south of the USA.

SKILLS INTERPRETATION

ACTIVITY

Look at Figure 9.20. Which state had the highest HDI? Find out why it had a higher value than the rest of India.

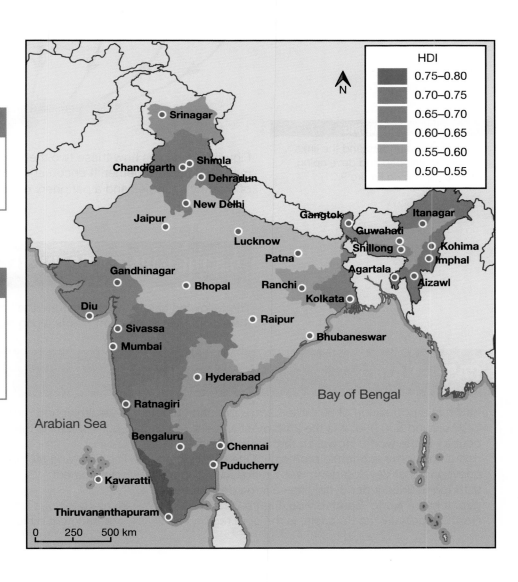

So, how can this internal unevenness in development be explained? It is helpful to return to the core-periphery idea hinted at on page 000. Very often economic development and wealth are concentrated in just one favoured region. It is generally the region containing the capital or leading city – the core. This concentration leaves other regions poor by comparison – the periphery. Core-periphery theory states that in the early stages of development cores grow at the expense of peripheries. But later on, economic growth and wealth slowly spread out or trickle down from the core. As a consequence, the periphery begins to pick up, and the **disparity** between the core and periphery becomes smaller.

▶ **Figure 9.21: Two interdependent worlds**

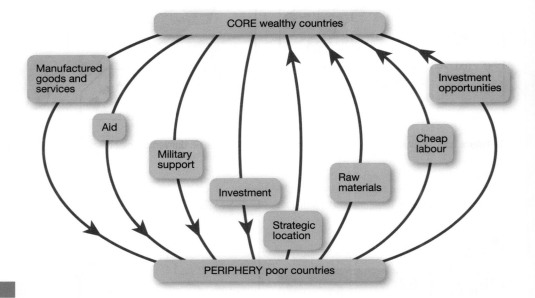

CHECK YOUR UNDERSTANDING

Check that you understand the links between developed and developing countries shown in Figure 9.21.

Figure 9.21 shows how these two theories (dependency and core-periphery) may be brought together. It encourages us to think globally in terms of a core of wealthy countries and a periphery of poor ones.

9.5 THE IMPACTS OF UNEVEN DEVELOPMENT

The following case studies illustrate the impacts of uneven development in three countries at different points along the development pathway.

CASE STUDY: ITALY – A DIVIDED COUNTRY

Italy has a population of 60 million and a per capita GDP around $30 000. It is the 26th most affluent country in the world. However, that figure hides the fact that it has some economic problems. One of these, an unevenness of development, was illustrated by Figure 9.18 (page 255). Indeed, Italy has a large development gap. It has a North–South divide (Figure 9.22). The North, especially the Po basin, is the core region and

is wealthier and more developed than the South. The South of Italy is the periphery.

The following factors contribute to the South's lesser development:

■ mountainous land makes communications and settlement difficult

- the climate of hot, dry summers and cold, wet winters is not ideal for agriculture
- the rocks are mostly limestone and form thin soils
- poor-quality grazing for sheep and goats
- poor transport links with the rest of country
- little employment outside agriculture – much emigration in search of work.

Since 1950 the Italian government and the EU have invested money to try to improve the South. As a result:

- some new autostradi (motorways) have been built
- new irrigation schemes allow tomatoes, citrus fruits and vegetables to be grown
- some large-scale manufacturing, such as iron and steel and production of motor vehicles, has relocated to the South.

Figure 9.23 shows that the Italian economy has been struggling since 2007. It also shows that the North–South divide remains and is getting wider. This is not surprising because in an economic recession it is the poorer areas that suffer more. The North continues to benefit from the factors that have made it the more prosperous part of Italy namely:

- good supplies of energy – natural gas in the Po basin and HEP from the Alps

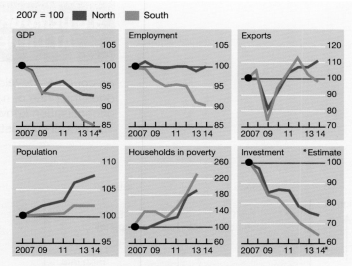

▲ Figure 9.23: Recent trends in Italy's North–South Divide

- more jobs in manufacturing and services; and a growing quaternary sector
- fertile lowland with irrigation water available
- large cities, for example Milan, Turin and Genoa (the so-called Turin industrial triangle) are connected by an efficient transport system
- close to large European markets
- better-quality housing and services, and a higher standard of living.

The South also receives many thousands of immigrants each year. These migrants risk their lives in sailing from North Africa across the Mediterranean, either as refugees or economic migrants.

SKILLS ▸ INTERPRETATION

ACTIVITY

Look at Figure 9.23. According to which measures has the gap between the North and South increased most?

SKILLS ▸ INTERPRETATION

ACTIVITY

Italy is one of a large number of countries with a 'North–South divide'. The UK is another. Looking at Figure 9.20 on page 255, do you think India is one?

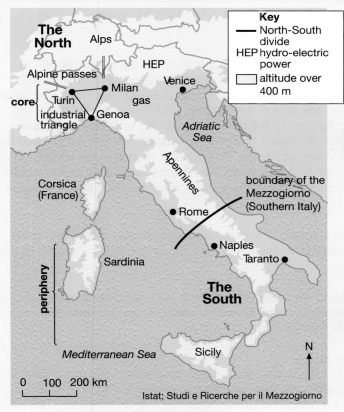

▲ Figure 9.22: The North–South divide in Italy

9.6 DEMOGRAPHIC CHARACTERISTICS OF COUNTRIES AT DIFFERENT LEVELS OF DEVELOPMENT

As countries develop, their birth and death rates change. As a result, so does their rate of natural increase in population. These changes underlie a generalisation known as the **demographic transition model** (Figure 9.24). The model suggests that countries pass through four or perhaps five stages.

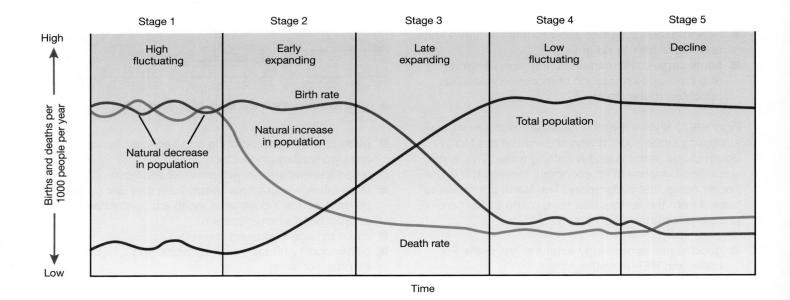

▲ **Figure 9.24: The demographic transition model**

As the rate of population growth changes, so does the population structure – the age and gender balance. These two important aspects of a population can be shown in a single diagram known as a **population pyramid**. The male population is shown on one side of the pyramid and the female on the other (Figure 9.25). The vertical axis of the pyramid is divided up into age-groups, usually of five years. The youngest group, 0 to 4 years, is at the bottom and the oldest group, over 90 years, at the top. The number of males and females in each age group is shown by a horizontal bar. The bar is proportional in length to either the number (as in Figure 9.25) or the percentage of males or females in the age group.

Let us now look at the stages of the demographic transition model and their typical population pyramids. At the same time, let us relate these stages to the development pathway.

Stage 1: High fluctuating – a period of high birth and death rates, both of which fluctuate. Natural change hovers between natural increase and decrease. Reasons for the high birth rate include:

■ little or no birth control
■ high infant mortality rate, which encourages couples to have more children
■ children may be seen as an asset and status symbol.

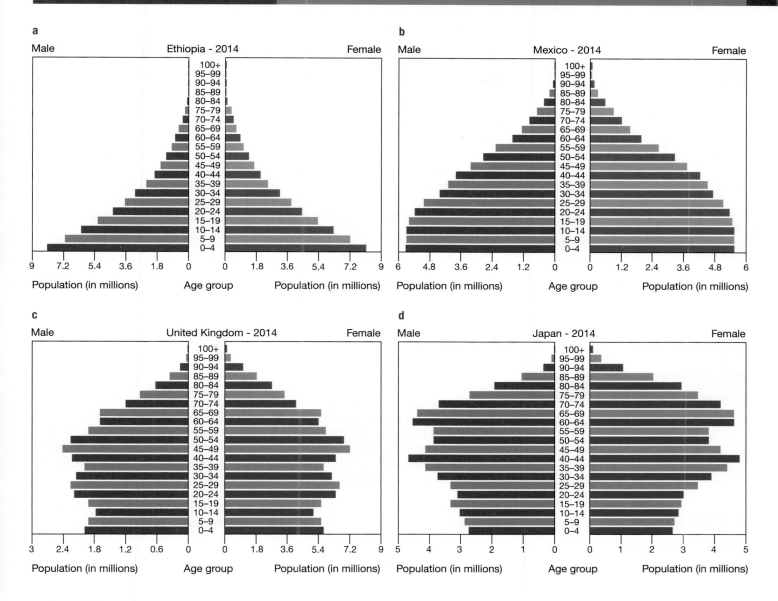

▲ Figure 9.25: Population pyramids for:
a) Ethiopia, b) Mexico, c) UK, and d) Japan

Reasons for the high death rate include:

■ high infant mortality
■ poor diet and famine
■ poor housing and hygiene, little or no healthcare.

Stage 2: Early expanding – a period of high birth rates, but falling death rates. The population beings to increase rapidly. Reasons for the falling death rate include:

■ lower infant mortality
■ improved healthcare and hygiene
■ better nutrition
■ safer water and better waste disposal.

The population pyramid for Ethiopia (a developing country just entering this stage) shows a youthful population (Figure 9.25a). The many young adults in the population are responsible for a high birth rate and many young children. Because the death rate is still high and life expectancy low, there are not many people aged over 55.

SKILLS INTERPRETATION

ACTIVITY

Write a brief description of how the population pyramids in Figure 9.25 change, starting with Ethiopia.

CHECK YOUR UNDERSTANDING

Check that you know the five stages of the demographic transition model.

SKILLS INTERPRETATION (STATISTICAL)

ACTIVITY

Identify the trends in the four demographic measures in Table 9.3.

SKILLS REASONING

ACTIVITY

Explain how and why development has an impact on human welfare.

Stage 3: Late expanding stage – a period of falling birth and death rates. The rate of population growth slows down as the rate of natural increase lessens. Reasons for the falling birth rate include:

■ widespread birth control
■ preference for smaller families
■ expense of bringing up children
■ low infant mortality rate.

The population pyramid for Mexico (an emerging country) has lost something of its pyramidal shape, but it is taller than the Ethiopia pyramid (Figure 9.25b). This means that the death rate is lower and life expectancy greater. The 'younger' part of the population is still larger than the 'older', but less so than in Ethiopia.

Stage 4: Low fluctuating – a period of low birth and death rates. Natural change hovers between increase and decrease. The death rate is kept low by improving healthcare. The birth rate is kept low by:

■ effective birth control
■ more working women choosing to delay the age at which they start having a family.

The population pyramid for the UK (a developed country) has almost lost its pyramidal shape because it bulges in the middle (Figure 9.25c). The base of the pyramid is undercut by the general decline in the birth rate. Interestingly, it looks as if the birth rate may be increasing a little. This possibly reflects the increasing number of young adult migrants entering the country. Even so, the 'older' and 'younger' components are more or less equal. The pyramid tells us that we have an ageing or greying population.

Stage 5: Decline – a period during which the death rate slightly exceeds the birth rate. The result is natural decrease and a decline in population. The population becomes even older because modern medicine is keeping elderly people alive for longer. This stage has only been reached by a few countries. The population pyramid for Japan (a developed country) is very undercut below the age of 40, but it is quite broad above the age of 65 (Figure 9.26d). This reflects a long life expectancy and overall makes the population pyramid almost top heavy. Japan's pyramid raises some interesting questions. Do populations continue to decline to a point where they disappear altogether? Or will immigration keep up the numbers, as in the UK?

	ETHIOPIA	MEXICO	UK	JAPAN
FERTILITY RATE (BIRTHS PER WOMAN)	4.4	2.2	1.8	1.4
DEATH RATE (PER 1000 PEOPLE)	8.2	5.3	9.3	9.5
INFANT MORTALITY RATE (DEATHS UNDER 1 YEAR OLD PER 1000 LIVE BIRTHS)	53.4	12.2	4.4	2.1
MATERIAL MORTALITY RATE (DEATHS PER 100 000 LIVE BIRTHS)	350	50	12	5

▲ Table 9.3: Demographic rates for four countries (2015)

It should be quite clear that the demographic characteristics of a country do change as it moves along the development pathway. Clearly, there are advances associated with development that have an impact upon populations. Particularly significant are better healthcare, education and hygiene, diet and housing.

9.7 STRATEGIES TO REDUCE UNEVEN DEVELOPMENT

There is a wide development and welfare gap between the richest and poorest countries. A number of governmental and non-governmental organisations are trying to narrow this gap by helping the poorest countries. These organisations include the United Nations and its various agencies such as the World Bank, the World Health Organization and UNESCO. Examples of non-governmental organisations include aid charities such as Oxfam, the Red Cross and Save the Children (Figure 9.26).

The best way to help the poorest countries is to help them develop. Economic development is important here. Economic growth creates the money needed to pay for education, housing and healthcare, which are such important aspects of quality of life. However, what is the best form for this help to take? We shall consider two common forms of help: aid and trade.

AID

▼ Figure 9.26: Two well-known charities associated with aid

▼ Figure 9.27: Appropriate aid, a basic water pump

Aid is the transfer of money, goods and expertise to assist developing countries and to help improve the quality of life. Aid can come from two sources. The governments of developed and emerging countries can contribute to two types of official aid. The first type is multilateral aid where a government donates mainly money to large international organisations such as the World Bank or UNESCO. These organisations then allocate that aid to those countries believed to be most in need.

The second type is bilateral aid where the government of a country gives the aid directly to the government of a receiving country. The aid may include grants of money, loans and technical help. Governments often give aid on certain conditions. The donor governments want something in return, perhaps a military base, the purchase of weapons or a trade agreement. For example, the USA gave Peru large amounts of aid to search for and exploit sources of oil. In return, Peru bought jet aircraft made in the USA and allowed fishing boats from the USA into Peruvian territorial waters.

As is explained below, aid in the form of loans is the least satisfactory type for the receiving country. It is much better if aid takes the form of technical assistance. However, the technology being transferred should be right for the circumstances of the receiving country. It is no good installing very sophisticated machinery if there are no people with the skills necessary to run and maintain that machinery. The challenge is to match the technology to the needs and skills of people in the receiving country. The answer is to be found in **intermediate technology** (Figure 9.27). This is a lower level of technology that is more accessible and useful to the people of a developing country. The technology needs to be easily understood. It should not require either a high level of training or the presence of foreign experts.

Voluntary aid is the aid provided by non-governmental organisations (NGOs) such as Oxfam and the Red Cross. Often the aid delivered by these NGOs takes the form of emergency aid delivered in response to some natural disaster such as an earthquake or drought (Figure 9.28 on page 262).

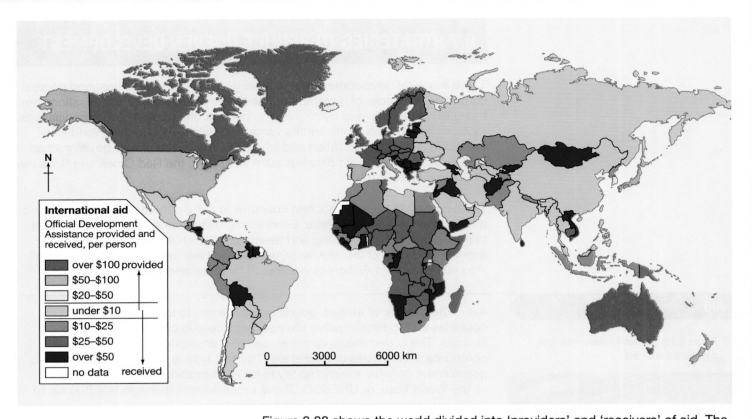

▲ **Figure 9.28: The global distribution of international aid (providers and receivers)**

SKILLS REASONING

Figure 9.28 shows the world divided into 'providers' and 'receivers' of aid. The division closely accords with the global North–South divide, but with at least two interesting exceptions:

■ Russia and the countries of Eastern Europe appear to be neither providers nor receivers
■ India and China are providers rather than receivers of aid (both countries provide aid with strings attached in their quest for resources).

ACTIVITY

Suggest some specific examples of appropriate aid.

TRADE

Trade is always thought to be a good way of stimulating economic development. In an era of globalisation, most countries want and need the chance to take part in international trade. Most countries have something which the rest of the world is prepared to buy. Those export sales allow a developing country to import what it needs to progress its economic development – machinery, vehicles, fertilisers, and so on. Unfortunately, world trade does not always operate fairly. There is much talk about free trade and ensuring that goods may be bought and sold across the world without duties, tariffs, quotas or import restrictions. The reality is that goods from developing countries often encounter various forms of **trade barrier**. The terms of global trade mostly favour the developed and emerging countries at the expense of the developing countries.

▲ **Figure 9.29: Some Fairtrade products**

The Fairtrade Foundation is a non-governmental organisation set up in 1992 to promote international trade. It seeks to obtain a fair price for a wide variety of goods exported from developing countries to the rest of the world (Figure 9.29). These goods include coffee, cocoa, sugar, tea, bananas, honey, wine and fresh fruit. The aim is to work with small-scale producers and help make them economically secure. This is an important step in encouraging economic development and eventually reducing global inequality.

SKILLS ▷ DECISION MAKING

ACTIVITY

Discuss in groups whether trade can ever be fair.

INTER-GOVERNMENTAL AGREEMENTS

We met some of the leading players in Part 8.2 – the World Bank, the IMF and the WTO. Most of these organisations are aimed at helping the poorest countries and ensuring that they are not discriminated against when it comes to international trade. Some specific examples are given in Part 9.9.

Debt relief involves developed countries no longer requiring developing countries to repay all their debts. Not everyone thinks that debt relief is a good idea. Some see it as giving out money to governments without any guarantee that they will spend it responsibly. Others say that the benefits are unlikely to reach the people most in need of help. Still others think that it will encourage corrupt governments to take on new debts.

9.8 CLOSING THE DEVELOPMENT GAP

Debt relief involves developed countries no longer requiring developing countries to repay all their debts. The core-periphery model suggests that a country's economic development will first gather around a core. The growth of this core will drain people from the rest of the country. So a periphery is created. However, there comes a point when the growth of the core becomes too large. It becomes congested and inefficient. This encourages a spread effect from the core and leads to new sub-cores of growth in the periphery. So eventually the periphery is revived and 'all is well'. Many governments have used this model as an excuse for doing little or nothing about closing development gaps. They claim that the problem will sort itself out.

Italy and the UK are among a number of countries that have made efforts to reduce their development gaps. The UK tried to slow down growth in the prosperous London and south-east England areas and encourage growth in the North. As in Italy, the approach was not successful. People have continued to move to the South. In Part 9.9 you can see that a new initiative is about to be taken.

Not all countries have their own development gaps, but all countries are aware of the global development gap. Everyone agrees that those on the wealthy side of the gap should help those on the poor side. Donor countries help either bilaterally or multilaterally (see page 261). But governments are agreed on two things:

■ that any help should be at a governmental level, that is from government to government
■ that the approach to closing the gaps should be top-down (see Part 9.9).

The approach of most NGOs, such as Oxfam, WaterAid and MSF, is to close the gaps from the 'grass roots level'. That is, to start by helping local people to help themselves. They claim that this is a more effective and immediate way of tackling the gaps. It is better than waiting for some government-led scheme.

9.9 CONTRASTING APPROACHES TO THE PROMOTION OF DEVELOPMENT

The previous section hinted at two different approaches to reducing the unevenness of development and to closing development gaps.

■ A top-down approach is government-led. It is usually concerned with the management of the national or regional economy as a whole. Governments and IGOs make the decisions with little involvement of local people. This approach often involves major infrastructural or industrial initiatives. The hope is that the development will eventually benefit regions and local areas.
■ A bottom-up approach is grass-roots based. It is centred on people and focuses on helping them to help themselves. It encourages the involvement of local people and benefits people at a local level. NGOs are often involved in such strategies, providing financial aid and technical expertise.

▶ Table 9.4: Development projects covered in this book

DEVELOPMENT LEVEL	TOP-DOWN	BOTTOM-UP
DEVELOPING	Ethiopia: Blue Nile Nigeria: coastal railway	Nepal: micro-hydro schemes Sri Lanka: urban gardening Kenya: WaterAid
EMERGING	China: SWNTP Thailand: DASTA	Brazil: Curitiba South Africa: agricultural co-ops
DEVELOPED	Spain: river management UK: HS2	UK: farm diversification

Table 9.4 shows examples of development projects that have been covered in earlier chapters, plus those that now follow.

Top-down is development usually funded and carried out by a government – either with the help of another government or of an inter-governmental organisation, such as the World Bank.

CASE STUDY: NIGERIA'S COASTAL RAILWAY

In 2014 the China Railway Construction Corporation (CRCC) signed a contract with the Nigerian government to build an express coastal railway line (Figure 9.30). The 1400-km route will link Lagos in the west to Calabar in the east. Crossing the huge delta of the Niger River, the line will also link the cities of Aba, Port Harcourt, Warri and Benin. The aim is to create a new coastal growth corridor that will encourage development right across the delta. The line will also link the two railway lines that serve the interior of Nigeria, one running north from Lagos and the other from Port Harcourt.

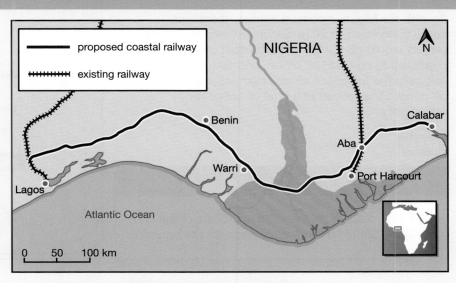

▲ Figure 9.30: The route of Nigeria's proposed coastal railway

The cost of the project will be nearly US$12 billion. It is predicted that it will create 200 000 jobs during the construction and around 30 000 permanent jobs once the line is opened. The funding arrangement is that 85 per cent will be provided in the form of loans from China and 15 per cent by the Nigerian government.

In 2016, with the fall in the global price of oil, the project appeared to be put on hold. The Nigerian economy is in recession. Oil exports are the mainstay of the Nigerian economy. But it is in Nigeria's best interest to have

SKILLS   REASONING

ACTIVITY

Explain how the building of this railway would allow Nigeria to become less dependent on oil.

the rail link built. It would give Nigeria the chance to become less dependent on oil.

CASE STUDY: THE UK'S HS2 PROJECT

High Speed 2 (HS2) is a planned high-speed railway that will link London, Birmingham, Leeds, Sheffield and Manchester. The first high-speed railway connected London to the Channel Tunnel and to Brussels and Paris. The first phase of the project will run from London to Birmingham (Figure 9.31). Construction is expected to start in 2017 and finish by 2026. At Birmingham the new rail link will divide with one branch running to Manchester and the other to Sheffield and Leeds. The building of these two

lines make up second phase and might be completed by 2033. From Manchester and Leeds there are links to Glasgow and Newcastle respectively but trains will run over existing slower tracks.

The project has been approved by the UK parliament. It will be funded by a mix of government money and private investment. HS2 is part of a wider government initiative to encourage a 'northern powerhouse'. It is yet another attempt to try to reduce the North–South divide in the UK. While the project has many supporters in the business and political worlds, there is much opposition. Some question whether it is worth spending all this money on what will be relatively small saving on present travel times. Others question the environmental costs associated with the building and running of the high-speed links.

As with the Nigerian coastal railway, there was some uncertainty in 2016 about whether or not the project would go ahead. With the cost currently predicted to be US$128 billion, and with the UK's exit from the European Union, perhaps it would be unwise to proceed at a time of economic uncertainty?

SKILLS ▸ DECISION MAKING

ACTIVITY

In class, discuss the benefits and costs of the HS2 project. Should the UK go ahead with the project?

▼ **Figure 9.31: The UK's HS2 routes**

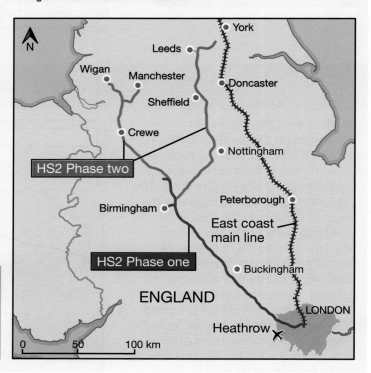

Top-down development schemes are usually very expensive and need heavy borrowing. Decision making is with the government and any external organisations that are involved. Local people are unlikely to have any say.

The problems with top-down development projects are most felt in developing countries for the following reasons.

- ■ The level of debt is often unsustainable.
- ■ Loans frequently have conditions attached, which result in 'non-nationals' gaining control over a country's affairs.
- ■ Often there is little benefit for local people other than perhaps employment during the construction phase. In this respect, the Nigerian coastal railway will be exceptional. The required machinery and technological expertise are imported.
- ■ The end product (railway, power station or irrigation scheme) is usually costly to operate and needs technical help with any maintenance.

BOTTOM-UP DEVELOPMENT

Bottom-up development schemes are projects that are planned and controlled by local communities to help their local area. They are not as expensive because they are smaller and use appropriate technology. Because of this, they are much less damaging to the environment than top-down schemes.

Bottom-up development is largely carried out and funded by NGOs. They work with local communities, teaching them how to help themselves. Often this will involve building services which they can repair and manage themselves. The approach is sustainable in the long run. It is also appropriate because it responds to the real needs of the communities.

CASE STUDY: WATERAID PROJECTS IN KENYA

Over 17 million Kenyans lack access to safe water. Over 32 million do not have access to adequate sanitation. These alarming figures have a huge impact on health and infant mortality.

Water is scarce in Kenya. At times, the country is on the brink of desertification. What little water Kenya has is not always distributed fairly. Priority is given to planned urban areas and to wealthy rural communities that can pay for it. This means that those people living in slums and remote villages often go without.

The sustainability of water, sanitation and hygiene programmes is also an issue. Although there is little reliable information, estimates suggest that up to a third of water pumps are broken at any one time.

WaterAid is a UK charitable trust set up in 1981. It seeks to help countries such as Kenya. WaterAid sets up partnerships with local communities to ensure that

SKILLS REASONING

ACTIVITY

Explain what rainwater harvesting is. What are its limitations?

the need for safe water and sanitation is met. They use technologies that are low-cost, appropriate to the local area and that can be easily maintained by the communities. Hand-dug wells, hand pumps and rainwater harvesting are commonly used to improve water supplies; community-managed latrines and wastewater management help with sanitation (Figure 9.32b).

The water and sanitation challenge in Kenya is huge. But NGOs, such as WaterAid, are beginning to make a difference as they enter into an increasing number of community partnerships.

▼ Figure 9.32: Meeting the need for: a) safe water, and b) sanitation

a

b

CASE STUDY: SOUTH AFRICA'S AGRICULTURAL CO-OPERATIVES

The Manyeding Agricultural Co-operative project in North Cape Province is one of a growing number of projects in South Africa fighting hunger and poverty in rural communities. These co-operatives are part of a government-backed scheme called 'End Hunger'. It seeks to promote self-sufficiency by helping communities to produce food – maize, beans, wheat, sunflower, ground nuts and potatoes – on communal and under-used land.

The initiative aims to help small-scale farmers to bring back into production 1 million hectares of land that has been lying fallow. It hopes to achieve this over a period of five years. It also aims to help small businesses process the crops once they have been harvested.

Financial support for the co-operatives is coming from the private sector. The Manyeding Co-operative, with 160 partners and 24 full-time employees, is funded by a local mining company (the amount is about R10 million) as well as by local municipalities. Some of this money is spent on agricultural machinery (Figure 9.34). A company, OrganiMark, is providing people working on the project with training both in organic farming and the marketing of produce. The provincial department of agriculture has provided money to build a package storing facility.

▶ Figure 9.33: Harvesting maize on the Manyeding Co-operative

The agricultural co-operative scheme is a neat one. Fallow land is used to grow food for consumption by co-operative members as well as for sale. Money from sales is providing much-needed income. Thus it is increasingly recognised that these co-operatives are winning the fight not just against hunger, but also food insecurity and poverty.

SKILLS ▶ CRITICAL THINKING

ACTIVITY

Write a brief account pointing out the strengths and weaknesses of this project.

CHECK YOUR UNDERSTANDING

Check that you understand the difference between a top-down and a bottom-up approach to development.

The four case studies have shown that both approaches to development have their benefits and costs. Bottom-up schemes, such as the South African agricultural co-operatives, help smaller communities to prosper; but they will not put right the system that makes them and millions of others poor and hungry. Top-down developments, such as the Nigerian coastal railway and the UK's HS2, may improve the national distribution of economic development. But will they really benefit the poorest people?

So, sometimes a bottom-up approach may be better; and sometimes a top-down approach is more beneficial. It all depends on the scale and seriousness of the uneven development. That is, whether the inequality is international or whether it exists within a country. A top-down approach is perhaps more effective when dealing with the global gap in economic development. However, the global gap in human welfare is better treated by a bottom-up approach. This is probably also true when dealing with development gaps within countries.

The important point here is that the two approaches are not mutually exclusive. The plight of the world's poorest countries and poorest people is so great that both approaches are desperately needed – preferably working in some sort of partnership.

CHAPTER QUESTIONS

END OF CHAPTER CHECKOUT

INTEGRATED SKILLS
During your work on the content of this chapter, you are expected to have practised in class the following:

- comparing the relative rankings of countries using composite development measures
- using and interpreting socio-economic data
- interpreting population pyramids
- using numerical economic data to profi le a country
- using and interpreting graphs showing rapid population growth.

SHORT RESPONSE

1 State **two** strands of development. [2]

2 Name **one** country with a large development gap. [1]

3 State **two** important components of population structure. [2]

4 State what the Gini coefficient measures. [1]

LONGER RESPONSE

8th **5** Explain why water has become such an important resource. [6]

9th **6** Explain how cores and peripheries develop. [6]

10th **7** Analyse the main factors contributing to human welfare. [12]

10th **8** Evaluate the costs and benefits of a top-down approach to the promotion of development. [12]

EXAM-STYLE PRACTICE

1 a) Identify **one** measure of economic development. [1]
   A Population density
   B Diet
   C Birth rate
   D Average income

b) State what the initials GDP stand for. [1]

2 a) Name the most widely-used measure of human welfare. [1]

b) Study Figure 9.24 (on page 258). Identify the stage of the demographic transition model that has been reached by most emerging countries today. [2]

c) Suggest **two** other distinguishing features of an emerging country. [2]

7th **3** Explain how **one** named inter-governmental organisation (IGO) is helping the growth of the global economy. [6]

4 a) Study Figure 9.28 (on page 262), Identify **two** countries that are leading providers of aid. [2]

b) Describe **two** possible problems facing countries receiving aid. [2]

7th **5** Compare **two** strategies for reducing international development gaps. [6]

10th **6** Analyse the main factors affecting the quality of life in the country where you live. [12]

[Exam-style practice total 35 marks]

GLOSSARY

abiotic non-living; a word describing some of the components in an ecosystem (e.g. rocks, climate)

abstraction removal of water from rivers, lakes and from below the water table

accessibility ease with which one location is reached from another

adjustment changes designed to react to and cope with a situation, such as the threat posed by a hazard

affluence wealth

agglomeration the concentration of people and their activities at particular locations

agro-forestry the growth of trees as windbreaks or as protection against soil erosion; also to provide fuelwood

alluvium the fine sediments deposited by streams and rivers

alternative energy renewable sources of energy, such as solar and wind power, that offer an alternative to the use of fossil fuels

aquifer permeable rock, such as limestone and sandstone, that is capable of holding and transferring underground water

arable farming a form of agriculture which concentrates on the growing of plant crops, such as cereals and vegetables

arch a coastal feature formed by the meeting of two caves cut into either side of a headland

asylum seeker a person who tries to enter a country by claiming to be a victim of persecution or hardship

bar a ridge-like accumulation of coastal sediments exposed at low tide

base flow the usual level of a river

bay a wide, coastal inlet that is open to the sea

beach an accumulation of coastal sediments, most often found in sheltered areas along the coast, such as bays

biodiversity the number and variety of living species found in a given area or ecosystem

biomass the total organic matter in plants and animals

biome a plant and animal community covering a large area of the Earth's surface

biosphere the parts of the Earth occupied by living organisms

biotic living or sustaining life; the term is often used to describe some of the components of an ecosystem (e.g. plants, animals)

birth rate the number of births in a population in a single year expressed per thousand people in that population

brownfield site land that has been previously used, abandoned and now awaits a new use

call centre an office used for receiving a large volume of incoming requests or enquiries by telephone; it may also be used in the opposite direction for telemarketing

carbon cycle the circulation of carbon between living organisms and their surroundings

carbon footprint the total amount of greenhouse gases produced as a result of human activities, usually expressed in equivalent tonnes of carbon dioxide (CO_2)

cash cropping growing crops for sale as distinct from crops grown for consumption by the farmer and his family

cave a hollow cut by the sea into the base of a cliff

channel network the pattern of linked streams and rivers within a drainage basin

clear felling the practice of cutting down all the trees on a site

cliff a steep rock slope, usually facing the sea

climate change long-term changes in global atmospheric conditions

closed system a set of interrelated objects in which there are no inputs and outputs

coastal cells a section of the coastline within which the movement of sediment is self-contained

commercial farming a type of agriculture in which production is intended for sale in markets

commodity chain see Production chain

communications the means or networks for moving information

commute the daily movement of people from their homes to their place of work

congestion acute overcrowding caused by high densities of traffic, business and people

conservation the protection of aspects of the environment for the future benefit of people

consumption the process of using goods and services to satisfy needs or wants

coral bleaching when sea water becomes too warm, corals expel the algae living in their tissues; this causes the coral to turn completely white

core the most important economic, political and social area of a country or global region – a centre of power

coriolis force the force created by the Earth's rotation that deflects any object moving at the Earth's surface

counter-urbanisation the movement of people and employment from major cities to smaller cities and towns, as well as to rural areas

cycle of poverty a self-perpetuating pattern of poverty and deprivation that passes from one generation to the next; it is perpetuated by poor educational opportunities and low incomes

dam a large structure, usually made of concrete, but sometimes made of earth, built across a river to hold back a large body of water (reservoir) for human use

death rate the number of deaths in a population in a single year expressed per thousand people in that population

decentralisation the movement of people, jobs and activities from the centre of major cities to the suburbs and beyond

deforestation the deliberate clearing of forested land, often causing environmental problems such as soil erosion

de-industrialisation the process whereby the importance of manufacturing in the economy declines

delta a landform produced by the deposition of sediment at the mouth of a river

demand the willingness or ability of consumers to pay for a particular good or service

demographic transition model a generalisation about how populations change over time, based on the experience of developed countries

depopulation a decline in the number of people living in a particular area

deposition the dropping of material being carried by a moving force, such as the waves

deprivation when the standard of living and quality of life fall below a minimum level

desertification the spread of desert conditions into what were previously semi-arid areas

development(1) making use of land for a variety of purposes, such as housing, industry, retailing and tourism

development (2) economic and social progress that leads to an improvement in the standard of living and quality of life for an increasing proportion of the population

development gap the difference in levels of development and standards of living between countries or regions

development indicators measures that are used to measure the level of development, such as per capita GNI and the literacy rate

diet the amount and kind of food consumed by a person or group of people

discharge the quantity of water flowing in a river channel at a particular location and time

disparity a great difference, for example between parts of a country in terms of wealth

disposable income income that is left after taxes and social security charges have been deducted; income that can be spent or saved as a household or person wishes

distributary a branch of a river or glacier which flows away from the main stream and does not return to it

dunes ridges of sand found in arid areas of the world and along coasts

earthquake a violent shaking of the Earth's crust

ecological footprint the impact a person or community has on the environment, expressed as the amount of land required to sustain their use of natural resources

ecosystem a community of plants and animals that interact with each other and their physical environment

economic migrant a person seeking work in another country

economic sector a major division of the economy based on the type of economic activity; the economies of all countries are made up of three sectors but most developed countries have a fourth sector

ecotourism a form of tourism that tries to minimise its impact on the environment; it is based on the use of local resources and labour, while its profits are enjoyed within the local area

emergency aid help in the form of food, medical care and temporary housing provided immediately after a disaster

energy heat and motive power; heat is provided by the Sun and by burning coal, oil and timber; motive power is provided by electricity, gas, steam and nuclear power

energy consumption the amount of energy used by individuals, groups or countries

energy gap a gap created because the loss of energy caused by phasing out the use of fossil fuels is greater than the amount of energy being developed from new, low-carbon sources

environmental quality the degree to which an area is free from air, water, noise and visual pollution

epicentre the point on the Earth's surface that is directly above the focus of an earthquake

erosion the wearing away and removal of material by a moving force, such as running water

estuary the mouth of a river as it enters the sea

ethnic cleansing when the actions of one ethnic or religious group force another such group to flee their homes, either by eviction or through intimidation and fear for their lives

ethnic group a group of people united by a common characteristic such as race, language or religion

ethnicity belonging to an ethnic group

evapotranspiration the release of moisture from the land by evaporation and transpiration by plants

extensive farming agriculture with low inputs of capital and labour associated with areas where land is cheap and plentiful; generally, yields are low per hectare

eamine a chronic shortage of food resulting in many people dying from starvation

fetch the distance of open sea over which the wind blows to generate waves

flood plain the flat land lying on either side of a river which periodically floods

flows steady streams of movement from one point to another

food chain (web) a series of organisms with interrelated feeding habits; each organism serves as food for the next in the chain

food miles the distance that food travels from the farmer who produces it to the consumer who eats it; the greater the distance, the greater the consumption of fossil fuels by transport and the greater the carbon emissions

foreign investment when a company or government becomes involved in the economy of another country

fossil fuel carbon fuels such as coal, oil and natural gas that cannot be 'remade', because it will take tens of millions of years for them to form again

fragile a term used to describe those natural environments that are sensitive to, and easily abused by, human activities

friction of distance the idea that the distance between places impedes movement between them; the greater the distance, the greater the friction or impediment

genetic engineering changing the genetic material of a plant or animal to produce, for example, a greater resistance to disease and a higher yield

genetically-modified (GM) crops plants used in agriculture whose DNA has been modified using genetic engineering techniques, usually to increase either yields or disease resistance

glasshouses structures of glass or polythene under which crops and flowers are grown; glasshouses protect the crops from wind and heavy rain, as well as maximising available sunlight

global economy the increasing economic interdependence of countries driven by the international spread of capitalism

globalisation a primarily economic process, increasing the integration of national markets for goods and services into a single global economy or market

global shift the movement of manufacturing from developed countries to cheaper production locations in emerging and developing countries

global warming the rise of global temperatures over time

gorge a deep, steep-sided and narrow valley usually occupied by a river

greenfield site land not used for urban development

greenhouse effect the warming of the Earth's atmosphere due to the trapping of heat that would otherwise be radiated back into space; it is vital to the survival of life on Earth

gross domestic product (GDP) the total value of goods and services produced by the economy of a country during a year

gross national income (GNI) the GDP of a country plus all the income earned by investments abroad

groundwater water contained within the soil or underlying rocks, and derived mainly from percolation of rainwater and meltwater

hard engineering protecting the coast by building structures such as sea walls and groynes

hazard an event which threatens the well-being of people and their property

headland an area of land jutting out into the sea

healthcare the provision of a range of medical services, such as clinics, hospitals, and homes for the elderly

human welfare the good fortune, health, happiness and prosperity of a person, group or organisation

hydrograph a graph showing the discharge of a river over a given period of time

hydrological cycle the global movement of water between the air, land and sea

immigration the movement of people into one country from another

industrialisation the process by which an economy is changed from a primarily agricultural one to one based on the manufacturing of goods

infant mortality children dying before they reach their first birthday

informal sector this comprises types of work that are not officially recognised; informal work is done by people working for themselves on the streets of cities, mainly in developing countries

information technology the use of computers and software to manage and process information

infrastructure the transport networks and the water, sewage and communications systems that are vital to people and their settlements and businesses

interlocking spur a series of ridges projecting out on alternate sides of a valley and around which a river winds

intermediate technology a technology that local people can use relatively easily and without much cost

irrigation the addition of water to farmland by artificial means, such as by pipelines and sprays

lahar a flow of wet material down the side of a volcano's ash cone which can become a serious hazard

land value the market price of a piece of land; what people or businesses are prepared to pay for owning and occupying it

levee a raised bank of material deposited by a river during periods of flooding

literacy the ability of a person to read and write

litter the layer of dead vegetation that over time is broken down by chemical and biological processes and incorporated into the soil

load the material (solid and dissolved) being transported by a river

long profile the section through a river or glacier course, from its source to its mouth

longshore drift the movement of sediments along the coast by wave action

malnutrition a condition resulting when a person has not eaten what is needed to maintain good health

mass movement the movement of weathered material down a slope due to the force of gravity

mass tourism popular, large-scale tourism of the type pioneered in southern Europe, the Caribbean, and North America in the 1960s and 1970s

meander a winding curve in a river's course

megacity a city or urban area with a population larger than 10 million

mitigation see Adjustment

natural disaster a natural event or hazard causing damage and destruction to property, as well as personal injuries and death

natural event something happening in the physical environment, such as a storm, volcanic eruption or earthquake

net migration the balance between the number of people entering and the number leaving a country or region

non-renewable energy energy produced from resources that cannot be replaced once they are used; examples include fossil fuels such as coal, oil and natural gas

offshore the seaward side of a coastline

onshore the landward side of a coastline

open system a set of interrelated objects in which there are both inputs and outputs

outsourcing a practice used by some companies to obtain goods or services by contract from an outside supplier, rather than providing those goods or services themselves

overgrazing when pasture or grazing is unable to support the number of animals relying on it for food; the result is that vegetation cover declines and soil erosion sets in

overpopulation when the population of an area cannot be adequately supported by available resources

ox-bow lake a horseshoe-shaped lake once part of a meandering river, but now cut off from it

package holiday a holiday in which travel and accommodation are put together by a tour operator and sold at a set and competitive price

pastoral farming the rearing of livestock for meat or other products such as milk and wool

paratransit transport services provided in developing countries by poor people and involving a range of small vehicles, such as rickshaws, which are available for public hire on a trip basis

periphery an area remote or isolated from the centre (core) of a country; it generally lags in terms of development and influence

permaculture a type of intensive agriculture that is both high-yielding and sustainable because it is based on, and takes advantage of, natural ecological processes

physical infrastructure the services, such as transport, telecommunications, water and sewage disposal, that are vital for people and business

plant succession the sequence of changes in the plant life of an ecosystem over time

plate movement mainly the coming together and the moving apart of tectonic plates

pollution chemicals, dirt or other substances which have harmful or poisonous effects on aspects of the environment, such as rivers and the air

population pressure when the number of people in an area exceeds the ability of that area to sustain them

population pyramid a diagram made up of horizontal proportional bars which shows the age and gender structure of a population

poverty where people are seriously lacking in terms of income, food, housing, basic services (clean water and sewage disposal) and access to education and healthcare. See also social deprivation

prediction forecasting future events or changes

primary sector economic activities concerned with the working of natural resources – agriculture, fishing, mining

production chain a sequence of stages in which companies exploit resources, transform them into goods and distribute them to consumers; it is a pathway along which goods travel from producers to consumers

push-pull factors the things that encourage people to migrate from one area to another; the negatives in the area of departure (push) are balanced against the positives of the destination (pull)

pyroclastic flow a devastating eruption of extremely hot gas, ash and rocks during a period of explosive volcanic activity; the downslope flow of this mixture is capable of reaching speeds of up to 200 kph

quality of life the degree of well-being and satisfaction felt by a person or a group of people

quaternary sector economic activities that provide highly skilled services such as collecting and processing information, research and development

raised beach a former beach now standing above sea level and some metres inland

refugee a person whose reasons for migrating are due to fear of persecution or death

renewable energy sources of energy which cannot be exhausted, such as the Sun, wind and running water

reservoir an area where water is collected and stored for human use

risk assessment judging the degree of damage and destruction that an area might experience as a result of a natural event

river regime the seasonal variations in the discharge of a river

river velocity the speed at which water is flowing in a river channel at a given location and time

rural environment areas of countryside largely concerned with farming and often with relatively low population densities

safe water water that is fit for human consumption

secondary sector economic activities concerned with making things, such as cars, buildings and generating electricity

sector shift change in the relative importance of an economy's sectors

selective logging the felling, at intervals, of the mature trees in a forest of mixed age; or the extraction of the most valuable trees from a forest

service provision making available commercial and social services, such as shops and schools

shanty town an area of slums built of salvaged materials; found either on the city edge or within the city, often on ground previously avoided by urban development

shifting cultivation an agricultural system in which a plot of land is cleared and crops are grown until the soil is exhausted; the plot is then deserted until the soil regains its fertility

slum a heavily populated urban area characterised by sub-standard housing and squalor; most lack basic services such as clean water, sanitation, electricity and law enforcement

social polarisation the process of segregation within a society based on income inequality and socio-economic status

soft engineering protecting the coast by working with nature

soil erosion the washing or blowing away of topsoil so that the fertility of the remaining soil is significantly reduced

sphere of influence (1) in geopolitics, an area over which a foreign power exercises control or influence

sphere of influence (2) an area over which a town or city distributes services or recruits labour and customers

spit material deposited by the sea that grows across a bay or estuary

squatter settlement see Shanty town

stack a detached column of rock located just off-shore and usually caused by the collapse of an arch

stakeholders individuals, groups or organisations that have an interest in a particular project or issue

stores features, such as lakes, rivers and aquifers, that receive, hold and release water

storm flow the increase in stream velocity caused by a period of intense rainfall

subduction the pushing down of one tectonic plate under another at a collision plate margin; pressure and heat convert the plate into magma

subsistence farming a type of agriculture concerned with the production of items to meet the food and living needs of the farmers and their families

suburbanisation the outward spread of the urban area, often at lower densities compared with the older parts of a town or city

supply making available something that is in demand

sustainable a term used to describe actions that minimise negative impacts on the environment and promote human well-being

system a set of interrelated components (stores) and processes (transfers). There are two types of system: an open system has inputs and outputs, and a closed system has none

tectonic plate a rigid segment of the Earth's crust which can 'float' across the heavier, semi-molten rocks below

temperate grassland a biome found in the dry interiors of continents with grasses rather than trees or shrubs

tertiary sector activities that provide commercial, professional and social services

tourism any leisure time or recreational activity which involves at least one night's absence from the normal place of residence

trade barrier a government-imposed restraint on the international flow of goods and services; the most common form of barrier is a tariff or tax on imports

trade bloc a group of countries drawn together by trade agreements promoting free trade between them

transfers the movement of water between stores in the hydrological cycle

transnational corporation (TNC) a large company operating in a number of countries and often involved in a variety of economic activities

transport the movement of a river's load

tropical cyclone a weather system of very low-pressure formed over tropical seas and involving strong winds and heavy rainfall (also known as a cyclone, hurricane or typhoon)

tsunami a tidal wave caused by the shock waves originating from an earthquake

underemployment a situation in which a person seeking full-time work can only find part-time jobs

urbanisation the process whereby a growing percentage of a population lives in towns and cities, and the economy depends increasingly on manufacturing and providing services

urban fringe the outer edge of a town or city when the built-up area gives way to the countryside

urban regeneration the investment of capital in the revival of older urban areas by either improving what is there or clearing it away and rebuilding

urban re-imaging changing the image of an urban area and the way people view it

urban re-branding developing a town or city to re-image it and change people's idea of it; promoting a town or city to a target audience or market

volcanic eruption the ejection of molten rock, ash or gases from a volcano

water balance the balance between inputs and outputs in the hydrological cycle or in a drainage basin

waterfall the vertical fall of a river's water, as the result of a band of hard rock running across the river channel

watershed the boundary between neighbouring drainage basins

wave a ridge of water formed by the circular movement of water near the surface of the sea

weathering the breakdown and decay of rock by natural processes, without the involvement of any moving force

well-being a condition experienced by people and greatly influenced by their standard of living and quality of life

world cities the world's leading cities, each being a major node in the complex economic, communication and decision-making networks produced by globalisation

INDEX

The author and publisher would like to thank the following individuals and organisations for permission to reproduce photographs:

(Key: b-bottom; c-centre; l-left; r-right; t-top)

123RF.com: dinozzaver 200tl/7.25, Kriangkrai Wangjai 105tl/4.9a, Mariya Smoliakova 200tr/7.24, Volodymyr Goinyk 139b/5.13c; **Age Fotostock Spain S.L.**: (c) VH / Oriental Touch 38br/2.11; **Alamy Stock Photo**: Alistair Laming 261c/9.26b, Andrew Findlay 28/1.35, Andrew Melbourne 109/4.15, Ashley Cooper 147tl/5.24a, Charles Sturge 224t/A, FLPA 206/7.33, Fredrick Kippe 245b/9.6, G&D Images 57b/2.34, Hemis 160/6.6, J Marshall - Tribaleye Images 23b/1.27c, JJ pixs 186b/7.4, Joerg Boethling 191t/7.11, John Warburton-Lee Photography 145b/5.22, keng po leung 170t/6.19, © Liu Peng / Xinhua 92/3.37, LOOK Die Bildagentur der Fotografen GmbH 224b/C, Manoj Attingal 176r/6.26, Marcelo Rudini 177b/6.28, Mike Goldwater 136l/5.8, mustbeyou CS 7, Peter Horree 147tr/5.24b, Plinthpics 236/8.22, Ramunas Bruzas 24c/1.29, REUTERS / Stringer 179/6.30, robertharding CS 8, Steve Morgan 103b/4.7, 200b/7.26, Susan Liebold 165c/6.13, Universal Images Group North America LLC 14b/1.17, 51/2.28, Victor de Schwanberg 153br/5.30; **AP Archive**: Rajesh Kumar Singh 153t/5.29; **Fairfax Syndication**: The Age / Penny Stephens 210/7.36; ©**Fairtrade www.fairtrade.org.uk**: 263/9.29; **Foster + Partners**: 176l/6.25; Getty Images: Arctic-Images 64, Buda Mendes / LatinContent 13tr/1.13, By LTCE 24b/1.30, Cem Genco / Anadolu Agency 225t/D, Chamila Karunarathne / Anadolu Agency 9/1.9, CS 10, Chamila Karunarathne / Anadolu Agency 9/1.9, CS 10, Dan Kitwood 86l/3.28, Dave Einsel 84b/3.25, DAVID J. PHILLIP / AFP 85t/3.26, DeAgostini / S. Vannini 46b/2.23, Dhiraj Singh / Bloomberg 105tr/4.9b, elisalocci CS 4t, Eye Ubiquitous / UIG CS 13, FLPA / Martin Hale 52t/2.29, Gavin Hellier / robertharding CS 4c, Geography Photos / UIG 104/4.8, Gregory Rec / Portland Press Herald 147br/5.24d, imagean 42/2.15, IndiaPictures / UIG 216t/8.3, Jake Rajs 14c/1.16, JIJI PRESS / AFP 79/3.18, 220/8.8, JIJI PRESS / AFP 79/3.18, 220/8.8, Jodi Cobb / National Geographic 170br/6.21, John van Hasselt / Corbis 154/5.31, © julhandiarso 78b/3.17, Kay Maeritz / LOOK-foto CS 14, LEON NEAL / AFP CS 11, Majority World / UIG 111b/4.18, Marka / UIG 81b/3.21, Markus Bsges / EyeEm 240, Michael Fay 182, Mike Coppola / Getty Images for Save The Children 261t/9.26a, NOAA 69t/3.5, Per-Anders Pettersson 225c/E, Philippe Marion 128, Photo12 / UIG 144r/5.21, Photos by W. Ebiko 32, Probal Rashid / LightRocket CS 2b, QAIS USYAN / AFP 178b/6.29b, RAPHAEL DE BENGY / AFP 66/3.1, Raquel Lopes / EyeEm 212, SIMON MAINA / AFP 149/5.26, SSPL 89/3.32, STR / AFP CS 12, Sunil Pradhan / Anadolu Agency 67tr/3.2c, Thomas Campean / anadolu Agency 267l/9.32a, Tim Graham 41b/2.14, Tim Graham Picture Library 147bl/5.24c, Tuul and Bruno Morandi 2, U.S. Navy 86r/3.29, ULTRA.F 156, VCG / VCG 25tr/1.32, VI Images 219b/8.7, Vicky_bennett 96, XiXinXing 225b/F; © Graham Crouch: 178t/6.29a; **NHPA Ltd / Photoshot Holdings**: © Xinhua 238/8.24; **Ordnance Survey**: CS 6br; **Panos Pictures**: Frederic Courbet 267r/9.32b; **Peter Facey**: CS 6tl, CS 6bl; **Shutterstock.com**: Aaron Gekoski 112/4.19, alexilena 38bl/2.10, alphaspirit 142r/5.18b, antb 140b/5.15, Artush 197b/7.21, Celso Diniz 180/6.31, ChameleonsEye 165t/6.12, Charles Harker 171/6.22, CroMary 224c/B, Danny E Hooks 195t/7.16, Ditty_about_summer 125b/4.32, drohn 101t/4.3, elenabsl 91b/3.32, Fitria Ramli 123l/4.28, fotosaga 195b/7.18, Franco Volpato 261b/9.27, Gaspar Janos 139t/5.13a, Gen Productions 139c/5.13b, Geniusksy 126tr/4.33, Gertjan Hooijer 41t/2.13, hecke61 108l/4.13, Infinity2 202/7.29, irakite 23t/1.27a, Janez Habjanic 268/9.33, Jo Chambers 37t/2.7, Joao Virissimo 134/5.6, John Wollwerth 172/6.23, Kingsly 158t/6.1, Kiwisoul 144l/5.20, Kodda 247/9.8, mary416 170bl/6.20, MC_Noppadol 237/8.23, MK Jones 110/4.16, Monkey Business Images 142l/5.18a, 242t/9.1, My Lit'l Eye 235b/8.21, Nair 85b/3.27, Ocean Image Photography 43t/2.16, Pablo Rogat CS 4b, Papa Annur 81t/3.20, Peter Turner Photography 52b/2.30, pupunkkop 187b/7.6, Richard Bowden 45b/2.21, Supermop 148/5.25, thiraphonthongaram 44b/2.19, thomaca 230/8.15, Timothy auld 122b/4.27, Vladimir Korostyshevskiy CS 3, Vladimir Melnikov 184/7.1, William Perugini 39b/2.12, Zhao jian kang 209/7.35; **University of Southampton Science Park**: CS 6tr

Cover photo/illustration © Getty images: wiratgasem

Inside front Cover: **Shutterstock.com:** Dmitry Lobanov

All other images © Pearson Education

eBook extra case studies:

Alamy Stock Photo: mustbeyou CS 7, robertharding 39t/2.12, CS 8; **Getty Images:** Chamila Karunarathne / Anadolu Agency CS 10, elisalocci CS 4t, Eye Ubiquitous / UIG CS 13, Gavin Hellier / robertharding CS 4c, Kay Maeritz / LOOK-foto CS 14, LEON NEAL / AFP CS 11, Probal Rashid / LightRocket CS 2b, STR / AFP CS 12; **Ordnance Survey:** CS 6br; **Peter Facey:** CS 6tl, CS 6bl; **Shutterstock.com:** Pablo Rogat CS 4b, Vladimir Korostyshevskiy CS 3; **University of Southampton Science Park:** CS 6tr